DATE DUE FOR RETURN

CANCELLED

16 JUN 1995

You

DYNAMICS OF STRUCTURAL SYSTEMS

DYNAMICS OF STRUCTURAL SYSTEMS

L.F. Boswell

and

C. D'Mello

The Department of Civil Engineering
City University, London

OXFORD

BLACKWELL SCIENTIFIC PUBLICATIONS

LONDON EDINBURGH BOSTON
MELBOURNE PARIS BERLIN VIENNA

© L.F. Boswell & C. D'Mello 1993

Blackwell Scientific Publications
Editorial Offices:
Osney Mead, Oxford OX2 0EL
25 John Street, London WC1N 2BL
23 Ainslie Place, Edinburgh EH3 6AJ
238 Main Street, Cambridge,
  Massachusetts 02142, USA
54 University Street, Carlton
  Victoria 3053, Australia

Other Editorial Offices:
Librairie Arnette SA
2, rue Casimir-Delavigne
75006 Paris
France

Blackwell Wissenschafts-Verlag GmbH
Meinekestrasse 4
D-1000 Berlin 15
Germany

Blackwell MZV
Feldgasse 13
A-1238 Wien
Austria

First published 1993

Set by Setrite Typesetters, Hong Kong
Printed and bound in Great Britain by
Hartnolls Ltd, Bodmin, Cornwall

DISTRIBUTORS

Marston Book Services Ltd
PO Box 87
Oxford OX2 0DT
(*Orders*: Tel: 0865 791155
         Fax: 0865 791927
         Telex: 837515)

USA
Blackwell Scientific Publications, Inc.
238 Main Street
Cambridge, MA 02142
(*Orders*: Tel: 800 759-6102
            617 876-7000)

Canada
Oxford University Press
70 Wynford Drive
Don Mills
Ontario M3C 1J9
(*Orders*: Tel: 416 441-2941)

Australia
Blackwell Scientific Publications Pty Ltd
54 University Street
Carlton, Victoria 3053
(*Orders*: Tel: 03 347-5552)

British Library
Cataloguing in Publication Data

A catalogue record for this book is
available from the British Library

ISBN 0-632-02199-3

Library of Congress
Cataloging in Publication Data

Boswell, L.F.
    Dynamics of structural systems/
    L.F. Boswell and C. D'Mello.
        p.   cm.
    Includes bibliographical references and
      index.
    ISBN 0-632-02199-3
    1. Structural dynamics. I. D'Mello,
    C.A. II. Title.
TA654.B64 1992
624.1'7—dc20   92-5312
                                   CIP

Contents

Preface

The aim of this book is to provide a sound understanding, both physical and mathematical, of the fundamentals of the theory of vibration and its application to practical problems, concentrating on structural problems. It is increasingly important for structural engineers to understand and solve dynamic problems, but the teaching of structural dynamics is relatively new to civil engineering degree courses, and the book will therefore be of use to those practising engineers whose formal studies did not include the subject. It is also suitable for undergraduates and postgraduates of civil engineering.

It has been assumed that readers have no previous knowledge of the fundamentals of vibration. However, some knowledge of solid mechanics and structural theory is required together with some appropriate mathematics. A second year engineering undergraduate will possess sufficient background knowledge.

In Chapter 1, the authors introduce the subject and emphasize the importance of dynamic modelling. The reduction of a practical engineering problem to a suitable dynamic model, from which equations of motion may be derived, is critical if meaningful results are to be obtained. Newton's second law and D'Alembert's principle for formulating the equations of motion are introduced in this chapter.

The various sources of dynamic loading are discussed in Chapter 2. It is important to identify a dynamic hazard and to have some knowledge of its properties. The chapter considers the dynamics of machines, impact events, blast, wind and earthquakes and assesses their effects. Many references have been provided. Although dynamic effects may not be damaging, they may have a nuisance value. Thus, consideration is given in this chapter to the assessment of levels of vibration and their effect on structures and humans.

Chapter 3 considers the behaviour of single degree of freedom systems and introduces much of the terminology to be used later in the text. The response of systems to harmonic, periodic, impact and general loading functions is considered. A section is devoted to the numerical solution of Duhamel's Integral for the response to general loading. A discussion of damping in structures is considered in some detail.

The objective of Chapter 4 is to introduce the reader to modern methods of dynamic analysis of multi-degree of freedom systems and to apply them to practical structural problems. Both lumped mass and continuous systems are considered. Exact matrix solutions are presented, such as mode super-position as well as approximate methods such as the Holzer and Stodola methods. Further approximate methods are presented for the analysis of continuous systems such as the Rayleigh and Rayleigh–Ritz methods.

Chapter 5 is devoted to a simple non-linear analysis of single degree of freedom systems subjected to a general loading function. Incremental equilibrium equations are derived and solved numerically. A computer program has been provided which the reader can implement if desired. This chapter should enable the reader to gain confidence in the difficult area of non-linear response and to interpret the response.

The final chapter is devoted to the measurement, recording and analysis of dynamic information. This is an important topic which contributes toward the solution of dynamic problems. In many instances, the structural engineer will have little knowledge of current sophisticated instrumentation. This chapter will provide a useful background and assist the engineer to communicate with the specialist. The chapter includes basic theory and principles of instrumentation and discusses typical instrumentation for measurement, recording and data analysis. Methods of data analysis are also discussed.

The authors would like to thank their respective families for allowing them the time to write this book. They would also like to thank a very patient publisher.

L.F. Boswell
C. D'Mello

Acknowledgements

The authors gratefully acknowledge the help and criticism of their colleagues in the Structures Division of the Department of Civil Engineering. They also acknowledge the work of numerous authors of dynamic texts whose material has influenced the style of the book. In particular, the authors are grateful to be allowed to reproduce some of this material.

The assistance of Mrs Pat Walker and Mr Kish Kundnani is acknowledged for their patient reproduction of the original script on the word processor.

Chapter One
Introduction to Structural Dynamics

1.1 Introduction

The objective of this book is to present a logical introduction and development of deterministic structural dynamics for the civil and structural engineer. Deterministic structural dynamics may be defined as that branch of applied mechanics which deals with the effects of sources of excitation which are fully prescribed and vary with time; the response of a structural system which is subjected to these sources will also be time varying. This book considers analytical and experimental procedures which may be used to determine the response of various structural systems. Although it is recognized that non-deterministic or random vibration theory represents an essential part of structural dynamics, the space limitations within this book make a satisfactory treatment impossible. However, occasional reference to random phenomena has been made in the text, where it has been considered relevant.

Time-varying excitations are usually referred to as sources of dynamic loading and may be represented by known or statistical assessed variations in the magnitude, direction or position of a force system. The associated structural response is usually described in terms of the variation of displacement with time and the associated stresses and strains are then obtained from subsequent calculations.

The majority of the structural systems which are to be considered are linear elastic and this implies that the principle of superposition is applicable. A later section is devoted, however, to the consideration of the non-linear behaviour of single degree of freedom systems.

1.2 Dynamic loading

Natural phenomena such as earthquakes or floods cause dynamic loading which can be catastrophic to all types of structures. Man-made sources of dynamic loading, however, are more likely to cause unserviceability due to

1

the disturbing effects upon people and equipment. Such forces can usually be controlled and remedial measures applied before structural damage occurs.

Codes of practice provide some information on certain types of dynamic loading in order to assist the engineer in defining an input motion. The engineer, however, is frequently required to conduct an investigation into a special source of dynamic loading which is of particular concern to the civil and structural engineer. A fuller account will be given in Chapter 2.

The simplest form of dynamic effect is the periodic loading which may be caused by out-of-balance machinery, and this may be represented by a sine wave. It is referred to as simple harmonic loading. The standard equations of harmonic motion may be used to describe this type of phenomenon. Established methods of measurement and analysis enable the maximum values and component frequencies of the motion of an installed machine to be determined.

Other types of periodic loading which might be encountered are more complex even though they are repeated in consecutive time periods. Again, these vibrations may be produced by machines or engine components. Such excitations may be represented by a Fourier series which consists of a sum of sine waves, the frequencies of which are multiples of the fundamental frequency. Both the simple and the complex forms of periodic loading may be treated in a similar manner.

There are many examples of loading which are not repeated in successive time intervals. These may be described as non-periodic and may be the source of vibration problems. Impulsive or short-duration excitation is characteristic of piling, shock and blasting operations and may have a non-zero value for a maximum time of only a few seconds. It may be represented by various equations depending upon the desired idealization. This may take the form of a sine pulse or a step pulse.

Other types of non-periodic loading may have a longer duration than shock or impulsive loads. The ground motion caused by a seismic disturbance for example is extremely complex and random.

The loads generated by wind and ocean waves are also the result of random processes. It is often acceptable to simplify the representations of these loads by retaining the main features of acceleration and frequency, for instance, and idealizing them deterministically.

Further sources of non-periodic dynamic loading which may be of concern to the engineer are caused by road traffic and railway vibration. Whatever the source of dynamic loading, the engineer must ensure that its representation is in a satisfactory form for subsequent analysis. Figure 1.1 illustrates some typical wave forms for periodic and non-periodic vibrations.

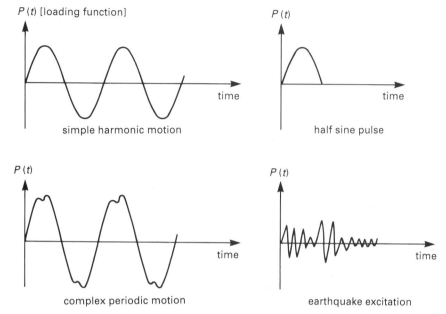

Fig. 1.1   Some examples of typical loading histories.

1.3   Structural idealization

Engineering structures are often very complex and in many cases it is impractical to include all the structural features in a dynamic analysis. Even with modern computers which are capable of solving very large problems there are still many situations where the causes and effects of vibration are so complicated that it is only possible to consider a simplified treatment in the design process. Each situation must be considered on its merits, but in every case the engineer must decide upon a suitable idealization to represent the structure. This is particularly important since it can determine the method of analysis and ultimately the size and complexity of the associated calculations.

Examples are provided throughout the text which will give an indication of the way in which certain structures may be idealized in order to permit a representative and rational analytical assumption. In this introduction, the general methods of idealization or discretization are considered with the associated methods of analysis.

The complexity of a structure may be defined by the number of dynamic degrees of freedom describing its motion. These degrees of freedom may be displacements, rotations, or some combination of both.

Consider the simple beam shown in Fig. 1.2 which supports two masses. Let it be assumed that initially each mass is constrained to move only in the

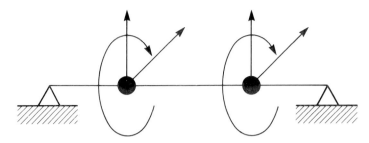

Fig. 1.2   A beam supporting two masses, the arrows indicate possible degrees of freedom.

vertical plane and that the total mass of the beam system is concentrated at the two points shown. For such an arrangement the system is said to have two degrees of freedom (2DOF). If each mass were allowed to move normal to the beam axis in the horizontal plane as well as in the vertical direction, there would then be a further two displacement components and the beam would have four degrees of freedom (4DOF). If each mass were allowed in addition to rotate about the beam axis, a further two rotational components would be added to the existing displacement components. The motion of the beam would then be described by six degrees of freedom (6DOF).

In general, the more complicated the structure, the greater the number of dynamic degrees of freedom which will be required to describe its motion. Clearly, the way in which a real structure can be simulated by a multi-degree of freedom system represents an important aspect of dynamic analysis.

1.4   Lumped mass idealization

One method of idealizing a structure is to concentrate the mass at various discrete points around the system. This type of representation is termed lumped mass idealization.

Consider the four-storey two-dimensional building frame shown in Fig. 1.3(a), which is assumed to be subjected to a horizontal ground motion. A reasonable assessment of the motion of the building suggests the movement of the structure will be dominated by the horizontal displacement of each storey. Furthermore, the main components of mass will be distributed between the four horizontal beams. If components of the total mass are lumped or concentrated at each storey and the structure is constrained to move only in the horizontal direction in the plane of the frame, then the dynamic behaviour of the structure is defined by the 4DOF lumped system shown in Fig. 1.3(b).

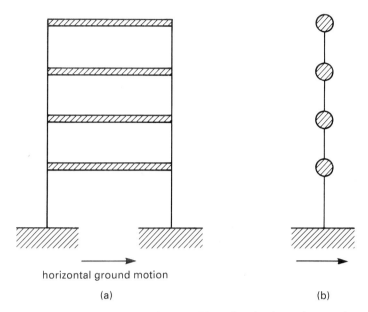

Fig. 1.3   (a) A four-storey structural frame subjected to horizontal ground motion. (b) A
4DOF lumped mass system which is dynamically equivalent to a four-storey
structural frame.

It is possible to reduce many types of structure to an equivalent lumped
system in a similar manner. In fact it may even be possible to reduce a
structure to a single mass system and still obtain useful results; this has the
advantage of making the analysis as simple as possible.

1.5   Distributed systems

The concept of concentrating the masses at discrete points in a structure is
not particularly applicable if the mass is uniformly distributed throughout.
Alternative methods of reducing the degrees of freedom for an acceptable
dynamic model may be used.

The simplest of these methods is to assume that the deflected shape of a
structure may be defined by the sum of a series of deflected shapes, which
become the displacement co-ordinates for the structure. For example, the
deflected shape of a simply-supported beam may be represented by the sum
of a sine wave series (Fig. 1.4). In principle, any series will suffice so long
as the beam boundary conditions are satisfied. The number of terms in the
series defines the number of degrees of freedom in the system.

An extension of this approach may be used in combination with the
lumped mass concept to analyse distributed systems, and is called the finite
element method. In this method, the structure is divided into elements,

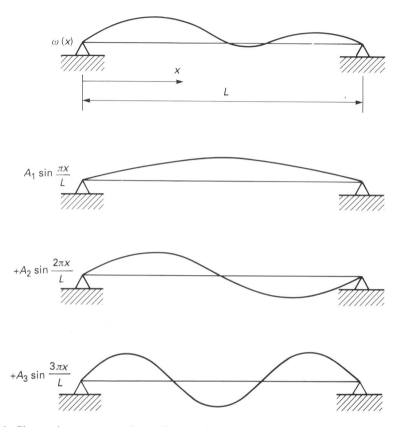

$\omega(x)$

x

L

$A_1 \sin \dfrac{\pi x}{L}$

$+A_2 \sin \dfrac{2\pi x}{L}$

$+A_3 \sin \dfrac{3\pi x}{L}$

Fig. 1.4 Sine series representations of the deflected shape of a beam.

which are connected at discrete points called nodes (Fig. 1.5). These nodes are allowed to displace and rotate in a prescribed manner to represent the motion of the structure. The sum of the displacements and rotations represents the total number of degrees of freedom for the system. The mass and other dynamic properties of the system are concentrated within each element. The complete structural system can then be assembled from the individual elements as part of the analytical procedure which is carried out using a computer.

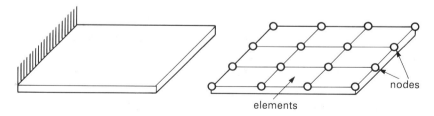

nodes

elements

Fig. 1.5 A finite element representation of a cantilevered plate.

The methods of structural idealization, which have so far been considered, reduce the dynamic behaviour of a structure to an acceptable number of degrees of freedom. To this extent they are, therefore, approximate methods of analysis. It is possible, by increasing the number of degrees of freedom representing the behaviour of a distributed or continuous system, to obtain results approaching those obtained from an exact solution. A truly exact solution, is, however, impractical since it can only be obtained by consideration of an infinite number of degrees of freedom.

Exact solutions for distributed or continuous systems may be obtained by solving the governing partial differential equations for the system. These define the equations of motion and problems involving beams, plates and shells may be studied by this means. Furthermore, the consideration of a structure as continuous rather than as a set of discrete points enables equations to be formed to study the effects of wave propagation in solids.

1.6 Formulation of equations of motion

Apart from a later consideration of the formulation of partial differential equations to describe exactly the equations of motion, the structural idealizations which are examined in this text are approximate and the formulation of the equations of motion may be obtained using either the principle of virtual work, D'Alembert's principle, or Hamilton's principle. These three methods which are used throughout the text will produce identical equations of motion of dynamic structural systems consisting of a representative number of degrees of freedom. Before these methods are discussed, consideration will be given to the physical nature of the dynamic problem. The linear elastic single degree of freedom (SDOF) mass–spring–damper system will be used to illustrate the essential features of the dynamic problem (Fig. 1.6).

The properties of this system are assumed to be defined and concentrated at discrete points. The mass of the system is signified by the symbol m, the elasticity of the spring stiffness k and the energy loss mechanism is represented by the viscous damping constant c. The latter is assumed to be proportional to the velocity of the motion. Additionally, a time-varying source of vibration $p(t)$ is applied to the mass.

All the physical components of the problem have now been defined. Since the loading source and the response of the system are both time-varying, the solution to the dynamic problem is not unique but is represented by a succession of solutions corresponding to time increments within the response history.

The excitation causes motion which is resisted first by the inertia of the mass due to its acceleration, secondly by the forces due to the elastic

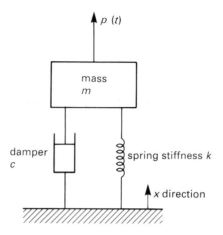

Fig. 1.6   Linear elastic SDOF system.

deformation, and thirdly by the energy loss mechanism. The elastic force is obtained as the product of the spring displacement and spring stiffness. The energy loss mechanism, which may be represented by a viscous damping force, is obtained as the product of the damper velocity and the viscous damping constant. Since the system has a single degree of freedom the displacement of the mass, spring and damper components will be the same. Similarly, the velocity and acceleration of the components of the system will be defined by single values.

Figure 1.6 will be used to explain and demonstrate the application of the three alternative methods of formulating the equations of motion for a dynamic system.

1.7 D'Alembert's principle

This principle enables the equations of motion to be obtained from a condition of dynamic equilibrium. Since Newton's second law states that the rate of change of momentum of a mass m is proportional to the force acting upon it, the following equation can be written for a force $P(t)$ acting upon a constant mass m

$$P(t) = m\,\frac{\mathrm{d}^2 x(t)}{\mathrm{d}t^2} = m\ddot{x}(t) \tag{1.1}$$

where $x(t)$ is the displacement of the mass and the dot notation represents differentiation with respect to time, i.e. $\ddot{x}(t)$ is the acceleration of the mass.

The product $m\ddot{x}(t)$ is the inertia force resisting the acceleration of the mass. The concept that this product resists motion is implied in D'Alembert's principle. The force $P(t)$ contains all the forces acting on the mass. In the case of Fig. 1.6, these forces are the elastic restoring force kx provided by the spring, the viscous damping force $c\dot{x}$ provided by the dashpot and the externally applied force $p(t)$ which is causing the motion. D'Alembert's principle allows the system being considered to be in dynamic equilibrium in which case $p(t)$, the applied force causing motion, is resisted by the inertia force of the mass, the spring elastic restoring force and the viscous damping force of the dashpot. The equation of motion for the SDOF system in Fig. 1.6 can be written accordingly as

$$m\ddot{x} + c\dot{x} + kx = p(t) \tag{1.2}$$

1.8 Virtual work principle

Equation (1.2) may also be obtained by the application of the principle of virtual work, which states that if a system of forces is given a virtual displacement consistent with the boundary conditions, then the total work done by all the forces within the system is zero. If the mass in Fig. 1.6 is given a virtual displacement δx, the forces in this system are displaced an identical amount and the total work done is given by

$$-m\ddot{x}\,\delta x - c\dot{x}\,\delta x - kx\,\delta x + p(t)\,\delta x = 0 \tag{1.3a}$$

$$[-m\ddot{x} - c\dot{x} - kx + p(t)]\,\delta x = 0 \tag{1.3b}$$

The negative signs take account of the fact that the associated forces act in the direction opposite to the virtual displacement. Since the virtual displacement is arbitrary and non-zero, Equation (1.3b) is identical to Equation (1.2).

1.9 Hamilton's principle

Hamilton's principle states that, if during an increment of time the variation of the kinetic and potential energy of a system is added to the variation of the work done by the non-conservative forces acting on the system, then the sum of these scalar quantities must be zero. The principle is embodied in the following equation

$$\int_{t_1}^{t_2} \delta(T - V)\,dt + \int_{t_1}^{t_2} \delta W\,dt = 0 \tag{1.4}$$

where T = kinetic energy of the system,
   V = potential energy of the system which includes the strain energy and the potential of external conservative forces,
   W = work done by non-conservative forces acting on the system which includes damping forces and external loads,
   δ = time variation.

Hamilton's principle will be used to obtain the equation of motion for the system in Fig. 1.6. The kinetic energy of the system is given by

$$T = \frac{1}{2} m\dot{x}^2 \tag{1.5a}$$

and the potential energy, which is simply the strain energy of the spring, is given by

$$V = \frac{1}{2} kx^2 \tag{1.5b}$$

The non-conservative forces in the system are the damping force $c\dot{x}$ and the applied load $p(t)$. The variation of the work done by these forces is the same as the virtual work expression associated with them and is given by

$$\delta W = p(t)\,\delta x - c\dot{x}\,\delta x \tag{1.5c}$$

Introducing Equations (1.5) into Equation (1.4), performing the variation of the first term and rearranging gives

$$\int_{t_1}^{t_2} [m\dot{x}\,\delta\dot{x} - c\dot{x}\,\delta x - kx\,\delta x + p(t)\,\delta x]\,dt = 0 \tag{1.6}$$

The term $m\dot{x}\,\delta\dot{x}\,dt$ is integrated by parts to give

$$\int_{t_1}^{t_2} m\dot{x}\,\delta\dot{x}\,dt = \left[m\dot{x}\,\delta x \right]_{t_1}^{t_2} - \int_{t_1}^{t_2} m\ddot{x}\,\delta x\,dt \tag{1.7}$$

where $\delta\dot{x} = \dfrac{d(\delta x)}{dt}$

In Hamilton's principle it is assumed that the variation δx vanishes at the limits of integration t_1 and t_2 and, therefore, the term in brackets becomes zero. If Equation (1.7) is substituted into Equation (1.6), the result may be written as

$$\int_{t_1}^{t_2} [-m\ddot{x} - c\dot{x} - kx + p(t)]\,\delta x\,dt = 0 \tag{1.8}$$

Since δx is arbitrary, Equation (1.8) is satisfied when the terms in the brackets equate to zero and Equation (1.8) reduces to Equation (1.2). This latter method is clearly more involved than the two previous approaches.

An essential step towards the solution of a dynamic problem is the establishment of the governing equations of motion. Examples are given throughout the text to demonstrate the application of the aforementioned methods of forming equations of motion for systems which have been idealised in a representative manner.

Chapter Two
Dynamic Loading, Its Assessment and Effects

2.1 Introduction

The initial stage in the solution of any vibration problem is to define the source of vibration. Information is required about the likely levels of forces, accelerations, amplitudes, etc., and the associated frequencies. There are many sources of vibration each producing their own particular characteristics. This chapter discusses the main sources of vibration which are of particular interest to the structural engineer. Even though it may have been possible to establish the likely levels of vibration, it is necessary to know whether these levels constitute a potentially damaging structural vibration or whether they are just a nuisance to the occupants of buildings. This chapter begins, therefore, with an assessment of vibration intensity. This will enable the engineer to assess the levels of vibration, which may have been obtained from field measurements, and to compare them with accepted criteria. The chapter then continues with a description of the various effects due to dynamic loading. A number of references have been provided in this chapter to facilitate further study of the subject material.

2.2 The assessment of structural vibrations

Increasing concern about the quality of the environment coupled with the development of urban and industrial areas has meant that many more vibration problems are encountered and demand more resolution than in the past. The engineer is, as a result, frequently involved in assessing the potential sources of vibration such as those resulting from, for example, railways, piling operations and industrial machinery. Such sources of vibration cause annoyance to the occupants of buildings and, in extreme cases, structural damage can occur.

While it is common practice in many countries to design structures to resist earthquakes, it is also becoming feasible to design structures which are isolated from lesser dynamic effects. Thus, multi-storey buildings have anti-vibration mountings incorporated within the foundations in order to eliminate the vibration and noise generated by underground railway systems beneath the structure (Waller 1965). In addition, parts of buildings have been completely isolated from the rest of the structure in order to eliminate annoying vibration effects (Morton 1967). It is clearly necessary to identify the factors which contribute to such vibrations, and to define criteria for the assessment of appropriate levels of human tolerance and structural damage. Such assessments are inevitably subject to variation and interpretation. There is, however, a certain amount of information available relating to human sensitivity, vibration intensity and the risk of structural damage. These areas, which will be discussed in some detail in this text, have applications in vibration survey work.

2.2.1 Human sensitivity

Although vibration effects may not cause structural damage they may be annoying to the occupants of buildings or they may affect the functioning of delicate instrumentation or manufacturing processes. Sufficient information is now available to enable an estimate to be made of the effect upon people of vibrations of known amplitude and frequency. The human body is extremely sensitive to vibration and amplitudes as low as 0.05 μm may be detected by the fingertips.

An important contribution to the assessment of human sensitivity was produced by Reiher & Meister (1931). In this work a group of people was subjected either to vertical or horizontal simple harmonic motion. Amplitudes and frequencies were measured and the associated sensations were noted. Vertical vibrations were found to be most easily detected when the subjects were standing, whereas horizontal vibrations were felt more readily by subjects who were lying down. The range of amplitudes and frequencies which were used in the experiments included those most likely to be encountered in structural vibrations.

By plotting amplitude against frequency, Reiher and Meister were able to classify vibrations into zones, the width of which allowed for subjective variation. Figure 2.1 shows the Reiher and Meister classification for vertical vibrations. For mild vibrations the intensity seemed to depend upon velocity, i.e. the sensation depending upon amplitude times frequency. For strong vibrations the intensity seemed to depend upon acceleration, i.e. the sensation depending upon amplitude times frequency squared.

A maximum velocity of approximately 0.04 in./s divides the 'just perceptible' zones; an annoying vibration corresponds to an approximate range of velocities from 0.1 to 0.3 in./s. The results produced by Reiher and Meister have been confirmed by other research workers.

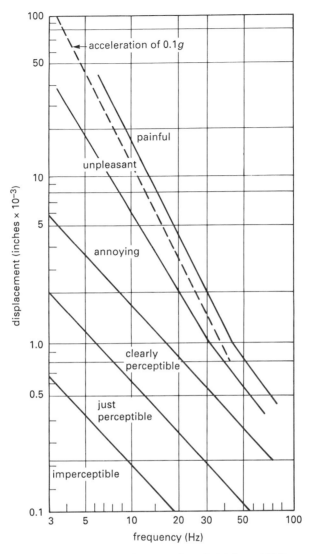

Fig. 2.1 Human sensitivity to vibration (after Reiher & Meister 1931).

Further information has been provided by Dieckmann (1958) and this has been used as the basis of the classification which is given in the German Standard DIN 4025 (1958). For vertical vibrations, Dieckmann has suggested that acceleration should be the criterion for frequencies below 5 c/s, velocity as the criterion for frequencies between 5 and 40 c/s and amplitude as the criterion for frequencies above 40 c/s. Dieckmann has suggested the same criteria for horizontal vibrations, but the frequency values are some-what different. For this case, the acceleration criterion corresponds to frequencies below 2 c/s, and the ranges for the velocity and amplitude criteria correspond to frequencies of 2 to 25 c/s and above 25 c/s respectively. Both Dieckmann and DIN 4025 define the level of vibration by a parameter K which is related as shown below to the previously stated criteria. Table 2.1 shows how the value of K is obtained and provides some typical values of the parameters which have been suggested by Dieckmann.

Table 2.2 gives the classification of the values of K which are provided by DIN 4025.

Some empirical formulae have been established to define the amplitude of vibration which produces discomfort. It has been suggested (Postlethwaite 1944) that the following expression can be used to define a discomfort amplitude

Table 2.1 Calculation of K and typical values (from Dieckmann 1958).

For vertical vibrations	For horizontal vibrations
Up to 5 c/s: $K = 25Af^2$	Up to 2 c/s: $K = 50Af^2$
From 5 to 40 c/s: $K = 125Af$	From 2 to 25 c/s: $K = 100Af$
Above 40 c/s: $K = 5000A$	Above 25 c/s: $K = 2500A$

A is the amplitude in inches and f is the frequency in c/s

K value	Description
0.1	Lower limit of human perception
1	Allowable in industry for any period of time
10	Allowable only for a short time
100	Upper limit of strain or endurance for the average man

Table 2.2 Classification of K values (DIN 4025).

K value	Classification	Effect on work
0.1	Threshold value. Vibration just perceptible.	
0.1–0.3	Just perceptible. Easily bearable, scarcely unpleasant.	Not affected
0.3–1	Easily noticeable. Bearable, but moderately unpleasant if lasting for an hour.	Still not affected
1–3	Strongly noticeable. Still tolerable, but very unpleasant if lasting over an hour.	Affected, but possible
3–10	Unpleasant. Can be tolerated for periods of up to 1 h, but not for longer.	Considerably affected, but still possible
10–30	Very unpleasant, cannot be tolerated for more than 10 minutes.	Barely possible
30–100	Extremely unpleasant. Not tolerable for more than 1 minute.	Impossible
Over 100	Intolerable	Impossible

discomfort amplitude, $x_0 = 0.076(1 + 194/f^2)$ mm   (2.1)

where f is the frequency. Dieckmann recommends the following alternative formula for discomfort amplitude.

$$x_0 = 0.076(1 + 125/f^2) \text{ mm} \tag{2.2}$$

The following expressions have been suggested for bridge design in order to avoid complaints of discomfort (Oehler 1957):

$$x_0 = 50.8/f^3 \text{ mm} \tag{2.3}$$

for $f = 1$ to 6 c/s, and

$$x_0 = 25.4/3f^2 \text{ mm} \tag{2.4}$$

for $f = 6$ to 20 c/s.

Table 2.3 compares some results which have been obtained from the above formulae.

The results of ten principal investigations into the prediction of levels of discomfort and annoyance have been summarized (Soliman 1963). These results are shown in Figs 2.2(a),(b),(c) and (d). It is worth noting that the quantity referred to as the rate of change of acceleration is also referred to as 'jerk'. This quantity has been found useful in defining vibration limits and may be obtained easily by differentiating the acceleration.

More recent information has been provided by Chen & Robertson (1972), Hansen *et al.* (1973) and Irwin (1978). The British Standards BS 6472 (1984) and BS 6611 (1985) are based on the work by Irwin (1978, 1981) British Standard 6472 is the standard for vertical vibration and defines thresholds of perception for different activities and types of building (Fig. 2.3 and Table 2.4). British Standard 6611 deals with low frequency horizontal motion of structures.

It will be seen in the following section that the levels of vibration which are likely to cause human discomfort are below those which would normally cause structural damage. The desirability to eliminate such vibrations from buildings is, however, still considered to be important in order to avoid annoyance to occupants.

Table 2.3 Amplitudes for discomfort vibration.

Authority	Amplitude (in.) at given frequency (cycles/second)			
	5	10	15	20
Postlethwaite	0.026	0.0088	0.0056	0.0045
Oehler	0.016	0.0033	0.0015	0.0008
Dieckmann	0.018	0.0068	0.0047	0.0039

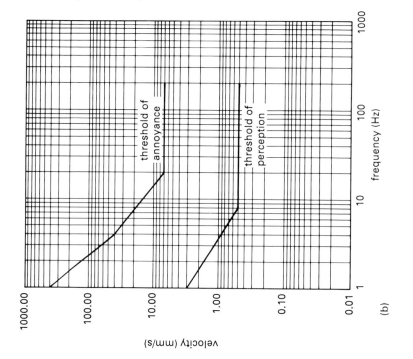

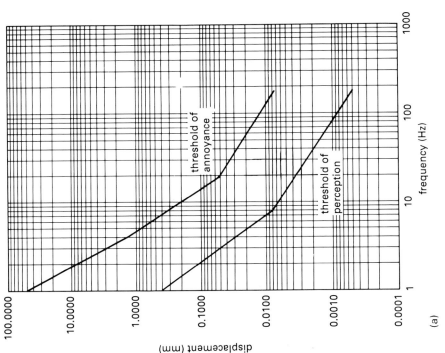

Fig. 2.2   Threshold of vibration perception and annoyance (after Soliman 1963). Frequency average values plotted against (a) displacement and (b) velocity.

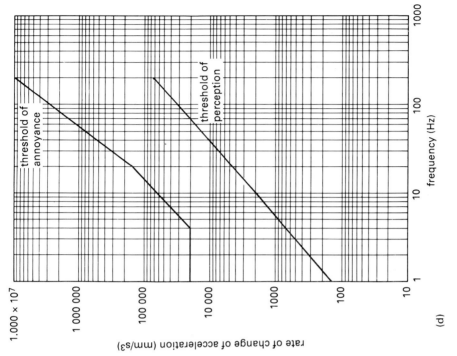

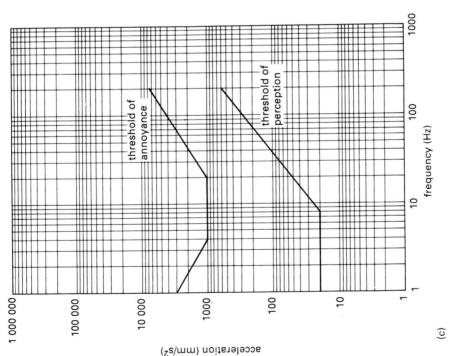

Fig. 2.2 (continued) Threshold of vibration perception and annoyance (after Soliman 1963). Frequency average values plotted against (c) acceleration and (d) rate of change of acceleration.

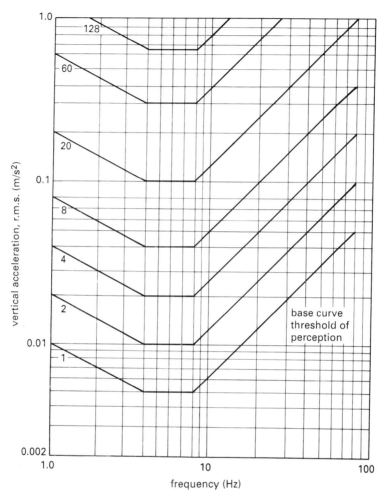

Fig. 2.3 Human response to vertical motion (BS 6472).

Table 2.4 Weighting factors above threshold of perception for acceptable building vibration (BS 6472).

Place	Time	Continuous or intermittent vibration and repeated shock	Impulsive shock with several occurrences per day
Critical working areas (e.g. hospital operating theatre)	Day	1	1
	Night	1	1
Residential	Day	2 to 4	60 to 90
	Night	1.4	20
Office	Day	4	128
	Night	4	128
Workshops	Day	8	128
	Night	8	128

2.2.2 Assessment of vibration intensity

A number of definitions have been proposed to assist in assessing vibration intensity. Most of the definitions are of German origin (Zeller 1931, 1933, 1941, 1949). The various definitions which are to be discussed in this section include Zeller's 'power' of vibration and scale of intensity; the 'damage figure'; the 'pal' unit and scale which were originally proposed by Zeller and the modified pal unit associated with the original German Standard DIN 4150 (1939). There would appear to be no internationally accepted standard to assess the intensity of vibration.

Considerations of energy and time form the basis of many of the units. The German sources define a basic energy unit, L, which considers the amount of kinetic energy involved when a unit mass vibrates for a time equal to one quarter of the period of vibration, hence

the energy unit, $L = 2\pi^2 A^2 f^3$ (cm^2/s^3) \hfill (2.5)

Zeller's power or strength of vibration which includes acceleration and frequency is defined by

Zeller's power, $Z = \dfrac{a^2}{f} = 16\pi^4 A^2 f^3$ (cm^2/s^3) \hfill (2.6)

There is a direct relationship between the energy unit, L, and Zeller's power, Z:

$$Z = 8\pi^2 L \simeq 80L \hfill (2.7)$$

It is possible to compare values of Zeller's power with the Mercentil–Sieberg–Cancani (MSC) scale of earthquake intensity which is discussed in section 2.4. Such a comparison has been given by Dawance (1957) and is shown in Table 2.5. The direct use of earthquake scales of intensity to predict the effects of industrial vibrations upon structures has been found to be in error since industrial vibrations usually have lower amplitudes and greater frequencies. The information provided in Table 2.5 corrects this anomaly where it can be seen that the earthquake and industrial vibration frequency ranges are used independently.

The vibrar unit of vibration intensity is used to compare vibrations of different frequency and amplitude and is derived from values of Zeller's power. The strength of a vibration in vibrar units is given as

strength (vibrar) $= 10\log_{10}(Z/Z_0)$ \hfill (2.8)

where Z is Zeller's power value (cm^2/s^3) and Z_0 has a value of $0.1\,cm^2/s^3$, thus

strength (vibrar) $= 10\log_{10}(10Z)$ \hfill (2.9)

Table 2.5   The Zeller scale and Z values.

Rating or grade	Assessment (i.e. effect on people or buildings)	Acceleration on Cancani scale (mm/s^2)	Zeller's value, Z (cm^2/s^3)
1	Not perceptible	1	1
2	Very light	2.5	2
3	Light	5	10
4	Measurable – small cracks (plaster cracks)	10	50
5	Fairly strong	25	250
6	Strong – beginning of danger zone	50	1 000
7	Very strong – serious cracking	100	5 000
8	Destructive	250	20 000
9	Devastating	500	100 000
10	Ruinous	1 000	500 000
11	Catastrophic	2 500	2 500 000
12	Very catastrophic	5 000	10 000 000

The graph shown in Fig. 2.4 enables values of the vibrar unit to be obtained from measured values of amplitude and frequency. The values which have been provided in this figure include those which are most likely to be caused by industrial sources of vibration. A strength of $10-20$ vibrar constitutes a light vibration, other ranges being given as: $20-30$, medium; $30-40$, strong; $40-50$, heavy; and $50-60$, very heavy.

A more commonly used unit of intensity is called the 'pal'. There are two alternative definitions for this unit. The original unit was suggested by Zeller (1933) and is defined by

$$\text{strength in pal (Zeller)} = 10\log_{10}(Z/Z_1) \tag{2.10}$$

where Z is the previously defined Zeller's power of vibration (cm^2/s^3) and Z_1 has the value $0.5\,cm^2/s^3$, thus

$$\text{strength in pal (Zeller)} = 10\log_{10}2Z \tag{2.11}$$

The pal (Zeller) definition is similar to the vibrar definition and the strength in pal (Zeller) is about seven units less than the corresponding strength in vibrar units for all frequencies. The pal (Zeller) scale of intensity is given in Table 2.6. Information has been produced by Koch (1953) which relates vibration intensity in pal (Zeller) units to sensation (see Table 2.7).

Note that the original classification considered an intensity of 70 pal as constituting the threshold of pain for vibrations of frequency greater than 15 c/s. For lower frequency vibrations in vehicles, ships, etc., this limit was taken as being at 55 pal.

The original German Standard DIN 4150 (1939) *Protection against*

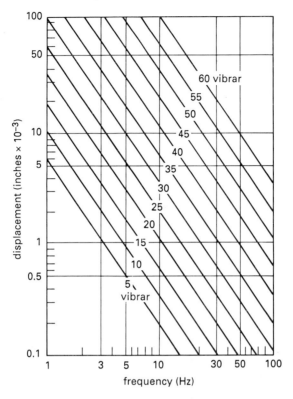

Fig. 2.4 The 'vibrar' unit of intensity.

Table 2.6 The pal (Zeller) scale of intensity.

Strength in pal (Zeller) units	Sensation or effect
0–10	Vibration perceptible, depending on body position.
10–20	General perception.
20–30	Traffic vibrations; not tolerable for persons in building.
30–40	Vibration in vehicles moving quietly.
40–50	Vibration in vehicles; accelerations in lifts.
50–60	Heavy vibrations in vehicles. Vibrations bearable by persons for short time without discomfort.
60–80	Physical discomfort; sea-sickness; pain, if associated with high frequencies.

vibrations of buildings defined the 'pal' unit in terms of velocity or energy ratios, i.e.

$$\text{strength in pal (DIN)} = 10\log_{10}(v_e/v_0)^2 = 20\log_{10}(v_e/v_0) \qquad (2.12)$$

where v_0 is a threshold velocity of $\frac{1}{10}\sqrt{10}$ cm/s and v_e is the root mean square value of the measured vibration velocity. This value may be obtained directly from standard measuring devices.

A vibration intensity of up to 5 pal (DIN) is just perceptible; 10 pal is clearly perceptible; 10–20 pal is severely perceptible or annoying and 40 pal is unpleasant.

The current German Standard DIN 4150, Part 2 (1986) on effects of people in buildings uses formulae to calculate perception intensity in the form of KB values. Acceptable levels are recommended for different situations and time of day.

It is important to be able to relate the results of a vibration survey to one of the accepted scales of intensity and to be consistent in order to avoid confusion between the various intensity scales which are available.

2.2.3   Assessment of structural damage

It is extremely difficult to state with any certainty the values of amplitude and frequency which are likely to cause structural damage. A realistic analysis of damage due to vibration requires the inclusion of a number of factors which might typically include the mass, stiffness and size of a building, the type of construction, and the fatigue properties of the structural materials. It seems that few measurements of stress have been made and reported in the literature at the same time as the measurement of amplitude and frequency. It should be possible, however, to make a reasonable estimate of the risk of damage if the various criteria for the assessment of vibration intensity are applied with some care and judgement. Comparison with previous situations in which damage may or may not have occurred is always worthwhile.

The most severe conditions to which a structure may be subjected are caused by earthquakes. For this case the values of horizontal acceleration and amplitude are relatively very large. Resonant motion is also a possibility since the natural frequency of a tall building, for instance, may be of the same order as the frequency of the main shock caused by an earthquake. Various seismic codes suggest that structures should be able to resist a horizontal acceleration of 0.1 times the gravitational acceleration.

The vibrar scale of vibration intensity has been used to assess the possibility of structural damage and Koch has produced some useful information which is given in Table 2.7 and Fig. 2.5.

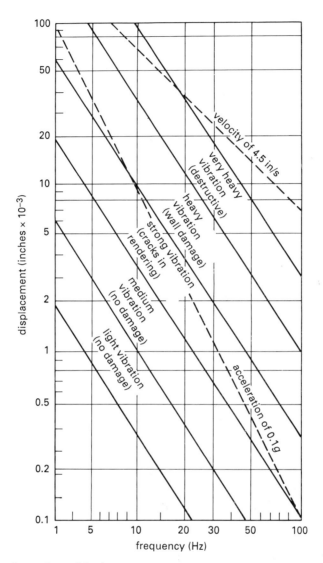

Fig. 2.5 Intensity and possible damage.

An alternative intensity unit called the 'damage figure' has been proposed by German researchers. The unit is expressed in mm^2/s^3 and is derived from the basic energy unit L, i.e.

damage figure $= 100L$ (2.13)

The 'damage figure' may be related to Zeller's power Z, i.e.

damage figure $= 1.266Z$ (2.14)

where Z has units of cm^2/s^3.

Table 2.7 Vibration intensity and probable damage (from Koch 1953).

Strength of vibration (vibrar)	Classification of vibration	Possible damaging effect
10−20	Light	None
20−30	Medium	None
30−40	Strong	Light damage (cracks in rendering, etc.)
40−50	Heavy	Severe (damage to main walls)
50−60	Very heavy	Destruction

Table 2.8 Damage figures and corresponding effects (from Koch 1953).

Damage figure (mm^2/s^3)	Equivalent strength in vibrar units	Extent of damage expected
50−500	26−36	Small cracks in rendering
500−2 000	36−42	Occasional slight cracks in walls
2 000−7 000	42−47	Cracks extend to main walls

Table 2.9 Recorded cases of severe vibration.

Case	Amplitude (in.)	Frequency (c/s)	Acceleration (g)	Damage figure, R (mm^2/s^3)	Strength (vibrar)	Estimated Reiher−Meister intensity
1	0.008 3	21	0.37	7 950	48	Painful
2	0.002 4	40	0.39	4 620	46	Painful
3	0.008	17	0.23	3 800	45	Painful
4	0.017 4	7.1	0.09	1 320	40	Unpleasant
5	0.001 3	29	0.11	520	36	Unpleasant
6	0.088	3.7	0.12	5 000	46	Unpleasant−Painful
7	0.052	6.2	0.20	8 100	48	Painful

Case 1. Measurements of vibration in a house in good repair, the outer walls being about 15 in. (381 mm) thick. The source of vibration is not stated but is referred to as 'short-duration' vibration. Cracks were produced in the outer wall, over door lintels, and also on ceilings and internal walls.
Case 2. Vibration of a building due to the operation of nearby billet shears. The brick wall, about 4.75 in. (120 mm) thick showed large through-cracks that later re-appeared after repair.
Case 3. Cracks produced in the outer walls of a house in good repair, due to the blowing up of rocks nearby.
Case 4. Machinery producing continuous vibration in the panels of a concrete floor. A through-crack appeared in one of the panels, and pieces of concrete became detached in a few places.
Case 5. Measurements of vibration in a house during explosions in a nearby quarry. Damage confined to a crack over a door lintel in a light partition wall.
Case 6. Measurements in a stone-built farmhouse when 200 lb. (90 kg) of gelignite was fired at a distance of 115 ft (35 m). Definite cracking of the plaster between partitions and ceiling.
Case 7. During dynamic tests to establish the sway frequency of a three-storey building, accelerations of the order 0.2g were reached, and in some cases maintained for almost a minute. No structural damage worse than a few hair cracks was sustained. Minute cracks were sometimes found to open and close completely, in phase with the exciter.

Koch has provided further information, which is given in Table 2.8, to enable vibration to be assessed as a potential cause of damage. By using the information which has been provided in Tables 2.7, 2.8 and Fig. 2.5, it is possible to use the measured values obtained from a vibration survey and to assess their damage potential.

Once more Koch has provided some useful information involving seven case studies of vibration damage. This information is given in Table 2.9 where it can be seen that the vibrations are unpleasant or painful according to the Reiher—Meister scale. It is useful to note that the various criteria for the assessment of damage correctly predict that damage should occur.

2.3 Dynamic effects due to wind

Probably the most common of all dynamic loading effects is that caused by wind. Multi-storey buildings and tall industrial structures such as stacks, towers and masts are subjected to lateral wind forces. The following dynamic effects which could be of relevance in practical design problems have been identified by Waller (1970):

(1) Gusting, the impinging of wind causing a dynamic load increment in addition to static loads.
(2) Buffeting which is a special case of gusting, where the stream of wind is modified by eddies, with some degree of periodicity, by the effect of another structure.
(3) Aerodynamic instability which is caused by vortices of regular pattern being generated in the lee of the structures. This situation is often referred to as vortex shedding.
(4) Galloping, which is a particular form of instability of slender structures. Large movements are associated with this instability; the lift and drag coefficients for some shapes vary with the angle of incidence of the wind causing movements from the position of equilibrium.

A considerable amount of research has been undertaken, usually in wind tunnels, to study the above dynamic effects. The overall variation of wind pressure is a random process. After an individual initial rise, the wind pressure in a gust may remain almost constant for about three or four seconds. Gradual increases in wind pressure give rise to peak effects which may be calculated statically. It will be seen in Chapter 3 that short duration shock loading causes additional loads of up to twice the equivalent static load depending upon the natural frequency of the structure.

A useful formula (Major 1980) may be used in a preliminary assessment of the dynamic load caused by wind. The dynamic magnifying factor is given by

$$\beta = \frac{\alpha}{\alpha^2 - 1} \sqrt{\left(1 + \alpha^2 - 2\alpha\sin\frac{\pi}{2\alpha}\right)} \tag{2.15}$$

where α is the ratio of the frequency of the wind force to the natural frequency of the building. The factor which is used to multiply the static load is equal to $\beta + 1$. For wind pressure which increases slowly and for relatively high natural frequencies, $\alpha = 0$ and $\beta = 0$. If the rise of wind pressure is sudden and the natural frequency of the building is low, $\alpha = \infty$ and $\beta = 1$, which is the worst condition. Rausch (1959) provides some typical values of β for different structures (see Table 2.10).

The concept of an equivalent static load should only be used for preliminary considerations. In general, references should be made to recommendations such as those given in the British Standards Institution Code of Practice, CP3: Chapter V: Part 2 (1972), concerning wind loading. According to CP3 an assessment of wind load should be made as follows:

A basic wind speed V appropriate to the district where the structure is to be erected is determined from meteorological data.

The basic design wind speed is determined from the formula

$$V_s = V S_1 S_2 S_3 \tag{2.16}$$

where S_1 is a factor related to topography, S_2 a factor which takes account of the combined effect of ground roughness, wind speed variation with height above ground and the size of the building under consideration, and S_3 is a statistical factor which takes account of the degree of security and period of time during which there will be exposure to the wind. Values of these factors appropriate for various conditions are provided in CP3: Chapter V: Part 2. A further directional factor S_4 is given in Appendix L of CP3.

The design wind speed is converted to the dynamic pressure q using

$$q = K V_s^2 \tag{2.17}$$

Table 2.10   Period of natural vibration and increment load for various structures.

	Stacks	Radio mast	Lighthouse	Tall building	Ordinary house
Period of natural vibration of various buildings (seconds)	2 to 3	1 to 4	1	1 to 5	2
Load factor β (%)	30 to 60	15 to 70	approx. 10	15 to 80	5 to 10

The values of K depend upon the units; for the SI system $K = 0.613$. CP3 provides tables from which it is possible to obtain the value of q for given values of V_s in three systems of units, i.e. N/m^2, kgf/m^2 and lbf/ft^2. The dynamic pressure q is then multiplied by an appropriate pressure coefficient C_p to give the pressure p exerted on any point on the surface of the building, i.e.

$$p = C_p q \tag{2.18}$$

The coefficient C_p may be negative; this indicates that p is a suction. The resultant load depends upon the difference in pressure on opposing surfaces and pressure coefficients C_{pi} and C_{pe} may be given for internal and external surfaces respectively. The resultant wind load is

$$F = (C_{pe} - C_{pi})qA \tag{2.19}$$

where A is the surface area.

An alternative method of obtaining the total wind load on a building is to use a force coefficient C_f, where one is available from design charts. For this case the total wind load is given by

$$F = C_f q A_e \tag{2.20}$$

where A_e is the effective frontal area of the structure. CP3 provides tables of pressure and force coefficients for a variety of building shapes. The ACI Committee 462 (1971) adopted a similar approach for the determination of lateral wind forces.

Vortex excitation may occur when a fluid in general, or the wind in particular, flows past an obstruction. Figure 2.6 illustrates the phenomenon for the case of a circular obstruction such as a tall chimney stack. The flow pattern depends upon the value of Reynolds number, $Re = vD/v$, where v is the fluid velocity, D is the cylinder diameter and v is the kinematic viscosity. For a range of Reynolds numbers the flow pattern consists of the alternate clockwise and anti-clockwise shedding of vortices in a 'vortex street' behind the cylinder. The frequency, f_n, at which the eddies are shed depends upon the Strouhal number, $S = f_n D/v$. The formation of vortices on either side of the cylinder gives rise to an alternating force which is

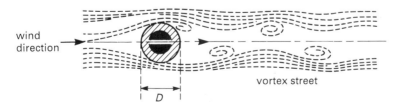

Fig. 2.6 Vortex shedding phenomenon.

transverse to the flow direction. Slender cylindrical structures are likely to be subjected to dynamic motions if their natural frequencies are similar to the frequencies at which the vortices are being shed. For structures possessing a circular cross-section the Strouhal number is approximately constant and equal to 0.2 for a practical range of wind velocities. If such a structure has a natural frequency of f_n, it is possible to calculate the wind velocity at which large vibrations may occur from the formula $v = f_n D/0.2$. The Strouhal number depends upon Reynolds number and on the shape of the cross-section which is obstructing flow.

Suspended cables or transmission lines may become subject to a phenomenon which is known as galloping and is observed as a large amplitude, low frequency unstable motion. The phenomenon is likely to occur when an elongated cross-section presents itself as an obstruction to wind flow. The formation of ice on a transmission line will cause an originally circular cross-section to become elongated with the long axis approximately parallel to the direction of wind flow. The transmission line is likely to become unstable in this orientation, since it is capable of absorbing energy from the aerodynamic force of the wind impinging on the cross-section. A treatment of this effect by Novak & Tanaka (1974), provides useful information on the subject.

The importance of wind tunnel testing cannot be over-emphasized. If a structure is of unusual shape or location it is recommended that experimental wind tunnel data be used in place of the coefficients provided in the various codes of practice.

A useful reference (Lehigh University 1972) deals with wind loading problems in different countries.

2.4 Dynamic effects due to earthquakes

The basic mechanisms beneath the earth's surface which cause earthquakes are not generally understood. For the present purpose it will be sufficient to assume that earthquakes are produced by breaks or dislocations in the earth's crust. An earthquake is initiated at some point beneath the earth's surface which is called the hypocentre. The epicentre is the point on the earth's surface which is directly above the hypocentre to the surface. The dilatational or P-wave and the distortional or S-wave are body waves which are recognizable from the seismographic trace. Since the P-wave has the greatest velocity it will arrive first at a measurement point. The Rayleigh and Love waves are two further waves which are propagated as a result of an earthquake, but travel only along the earth's surface. Figure 2.7 shows the horizontal ground acceleration recorded at El Centro, California during the 1940 earthquake.

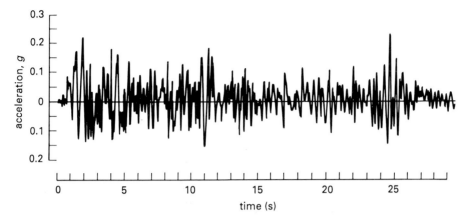

Fig. 2.7   The east−west component of horizontal ground acceleration recorded at El Centro, California, during the earthquake of 18 May 1940, centred approximately 30 miles (50 km) away horizontally and 15 miles (25 km) beneath the surface of the ground. The north−south component of horizontal ground acceleration was similar but slightly more intense with a maximum acceleration of 0.33 g. The vertical component had accelerations approximately 60% as large as the horizontal components and had markedly higher frequency components.

The ground motion which is associated with an earthquake is very complex with components of motion in both horizontal and vertical directions, the motion in the horizontal direction being of particular importance to the response of structures. In order to establish the intensity of an earthquake various methods of classification are available. A minor earthquake which may be considered as a shock, will cause ground accelerations of the order of 0.1 times the gravitational acceleration (0.1 g) whilst strong motion random earthquakes cause ground accelerations approximately within the range 0.1 g to 1.0 g.

For general purposes earthquakes are usually classified by their magnitude in one of the alternative intensity scales. Of these, the Richter scale (Richter 1935) and the modified Mercentil−Sieberg−Cancani (MSC) scale (Richter 1958) are probably the best known. Richter expressed the intensity of an earthquake in terms of the logarithm to the base 10 of the maximum amplitude in microns on a standard short period seismograph at a distance of 100 kilometres from the epicentre. Observed amplitudes at other distances may be reduced to the standard distance by the use of tables. The intensity of an earthquake may also be expressed in terms of energy. On a Richter scale, minor earthquakes would correspond to a magnitude of 1.5 and catastrophic earthquakes to a magnitude of 8.5. There are eight scales of intensity or magnitude in the Richter classification.

The MSC classification has twelve scales which correspond to different values of acceleration (see Table 2.11). Table 2.12 (Major 1980) compares the Richter and MSC scales of earthquake intensity.

Table 2.11   MSC earthquake intensity scale.

Rate of intensity	Maximum acceleration		Maximum velocity	Maximum mean amplitude value (mm)
	(g)	$(cm\,s^{-2})$		
VI	0.005–0.010	5.00–10.00	1.00–2.00	0.50–1.00
VII	0.010–0.025	10.00–25.00	2.00–4.00	1.00–2.00
VIII	0.025–0.050	25.00–50.00	4.00–8.00	2.00–4.00
IX	0.050–0.100	50.00–100.00	8.00–10.00	4.00–8.00

Table 2.12   Comparison of the Richter and MSC scales (from Major 1980).

Richter scale (magnitude)	Energy (erg)	MSC scale
3.00–3.90	$9.5 \times 10^{15} - 4.0 \times 10^{17}$	I–III
4.00–4.90	$6.0 \times 10^{17} - 8.8 \times 10^{18}$	IV–V
5.00–5.90	$9.5 \times 10^{18} - 4.0 \times 10^{20}$	VI–VII
6.00–6.90	$6.0 \times 10^{20} - 8.8 \times 10^{21}$	VII–VIII
7.00–7.90	$9.5 \times 10^{22} - 4.0 \times 10^{23}$	IX–X
8.00–8.90	$6.0 \times 10^{23} - 8.8 \times 10^{24}$	XI–XII

From the earthquake resistant design viewpoint it is necessary to specify the design earthquake which is usually associated with the ground motion causing a structure to respond critically. In practice, this cannot be defined exactly since the ground motion associated with predicted earthquakes can only be estimated and the critical response will vary according to the different limit states controlling design specifications.

In cases where serviceability limit states govern the structural design, structures should remain essentially elastic during the earthquake motion. Although peak site acceleration has been used in the past to define a design earthquake, it is now generally accepted that one of the most satisfactory ways to describe quantitatively the serviceability level design earthquake is to use an average response spectrum. A velocity spectrum is obtained from a SDOF system subjected to earthquake motion. The maximum velocity is a function of the natural period and the fraction of critical damping in the system. The velocity response spectrum represents a plot of the maximum velocity against the natural period of vibration for specified amounts of damping (Fig. 2.8). Thus, it is possible to obtain values of acceleration, velocity and amplitude for different values of damping at different times in the time-history of the earthquake. Response spectra are discussed further in Chapter 3, section 3.6.

The average spectrum is obtained from a statistical analysis of the linear elastic response spectra of ground motion records resulting from earthquakes with comparable magnitudes and obtained at sites with similar epicentral distances and soil conditions to the proposed location. The procedure has

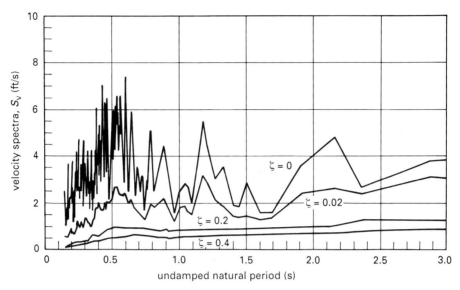

Fig. 2.8 Velocity spectrum for the east−west component of the earthquake recorded at El Centro, California, 18 May 1940. The fraction of critical damping is indicated by ζ.

been used (Newmark *et al.* 1973) to predict possible design response spectra.

The maximum response of a mode of vibration of a more complex system may also be determined from a response spectrum.

If safety rather than serviceability considerations control the structural design then allowable inelastic deformations can be permitted. Most structures have a considerable capacity for energy dissipation and advantage may be taken of this property. The non-linear dynamic analysis of practical engineering structures subjected to earthquake motion is complex and there does not seem to be a generally accepted form which may be suitable for design. Preliminary design loads can be obtained from inelastic design response spectra which are obtained from the non-linear dynamic response of structural models with realistic non-linear properties. An indication of this procedure is given in Chapter 5 which considers non-linear response. Simpler methods, which have been considered by Newmark (1978), use modified linear elastic response spectra to represent inelastic response spectra.

The lateral forces which are induced in the structure by the earthquake ground motion are obtained from the design earthquake spectrum. The structure may then be designed to resist these forces.

Instead of obtaining loading from a direct dynamic analysis using the design earthquake, it is possible to obtain the lateral forces acting on a tall building from recommendations in codes of practice such as the ACI Committee 442 (1988) recommendations and the *Recommended lateral force requirements and commentary* (1973), both of which are used in the USA. In these cases, the effect of earthquake loading is represented by the maximum shear force acting at the base of a building and is given by $V_{max} = KCW$. The weight of the building is represented by W; C is a base shear coefficient and is given by $C = 0.05/^3\sqrt{T}$, where T is the fundamental period of vibration of the structure. The factor K depends upon the form of construction and varies from $\frac{2}{3}$ for rigidly jointed frames to $\frac{1}{3}$ for shear wall construction. The fundamental period of vibration is obtained from the approximate expression, $T = cH/\sqrt{B}$, where H is the height of the building and B the length of the side which is parallel to the direction of motion. The coefficient c depends upon the foundation and values are given in the references of the ACI and the Structural Engineers' Association.

The maximum base shear force V_{max} can be distributed throughout the building at each storey level according to the formula

$$f_{imax} = \left(\frac{w_i h_i}{\Sigma w_i h_i}\right) V_{max}$$

where f_{imax} is the maximum lateral force at level i, w_i is the weight at level i, and h_i is the height of level i above the base of the building.

The Applied Technical Council of the National Bureau of Standards, USA (1978) have also provided recommendations for the seismic design of buildings.

2.5   Dynamic effects due to machinery

Excessive dynamic effects caused by machinery may be potentially undesirable for two reasons. They may indicate the improper functioning of the machine or the machine-foundation system, or they may generate vibrations which are deemed to be a nuisance or even damaging. Limits on the levels of machine vibrations are, therefore, considered to be necessary criteria for the design of machines and their foundations. For steady-state vibrations, limits on the values of amplitude, velocity or acceleration are usually provided to ensure satisfactory operation.

If these values are either calculated or measured at the machine bearings, they may be compared with various charts and tables for assessment.

Graphs of vibration amplitude versus frequency have been provided for different machine types. Figures 2.9(a) and (b) show the information provided by Rathbone (1939) for heavy machinery and Buzdugan (1968) for electro-motors of the Schenck−Darmstadt type.

The VDI recommendations (1941) appear to be the most generally applicable. These recommendations classify into the following four groups and provide information for each group.

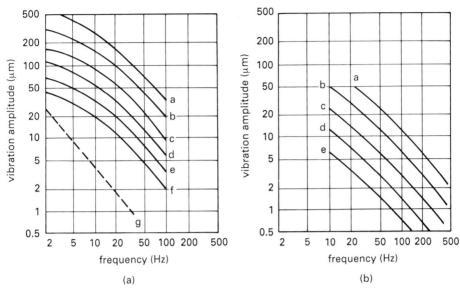

Fig. 2.9   (a) Charts for vibration tolerances for heavy machinery. a − very rough, b − rough, c − slighly rough, d − fair, e − good, f − very smooth, g − limit of perceptible vibration. (After Rathbone 1939.) (b) Charts for electro-motors of the Schenk−Darmstadt type. a − very rough; b − rough, c − slightly rough, d − admissible, e − good to very good. (After Buzdugan 1968).

Group K: Individual parts of driving gears of prime movers and machine tools, connected finally in operation to the rest of the machine, including mass-produced electro-motors meeting no special requirements and being only up to 15 kW (Fig. 2.10(a)).

Group M: Machines of medium size, i.e. electric motors of 15 to 75 kW, without the requirements of special foundations, and rotating machinery up to 300 kW placed on foundations (Fig. 2.10(b)).

Group G: Major parts of driving gears mounted on rigid or heavy foundations highly overtuned in the direction of the vibration considered; large prime movers and machine tools having only revolving masses (Fig. 2.10(c)).

Group T: Large prime movers and machine tools with only revolving masses, supported on special foundations, deeply undertuned in the direction of the vibration considered, e.g. turbine sets, especially those with modern, light foundations (Fig. 2.10(d)).

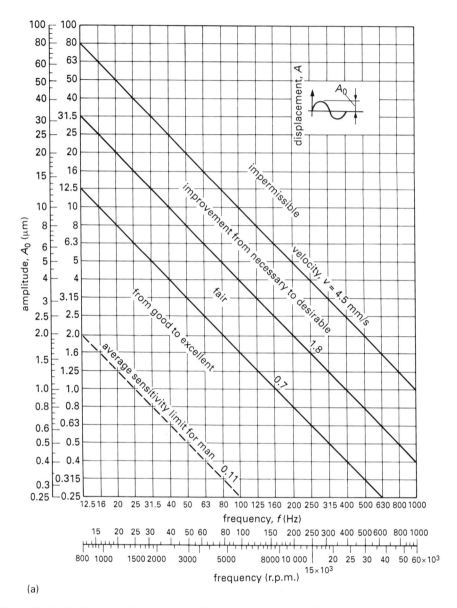

Fig. 2.10 VDI directives for assessing vibrations: (a) Group K.

In order to use the VDI recommendations, the amplitudes and velocity should be measured at machine bearings.

Korchinski (1948) has provided some useful information which is shown in Table 2.13. An indication of permissible amplitudes for various types of machines may be obtained from this table.

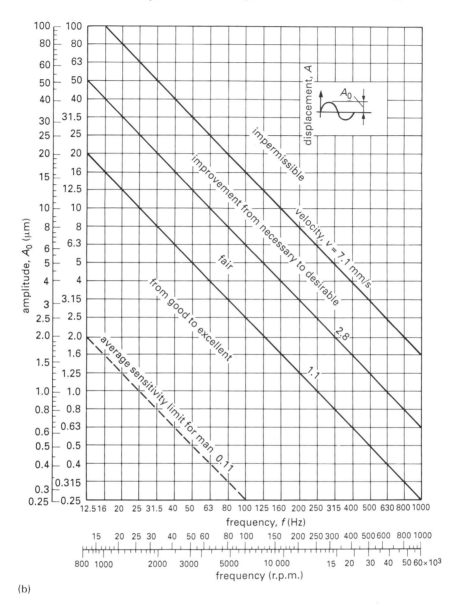

(b)

Fig. 2.10   (continued) VDI directives for assessing vibrations: (b) Group M.

The British Standards Institution BS 4675 Part 2 (1978) has made recommendations for the comparative evaluation of vibrations for four classes of machinery (Table 2.14). Qualitative judgements, A to D, are made for each class and these may be compared with previously mentioned criteria for perception and annoyance, etc.

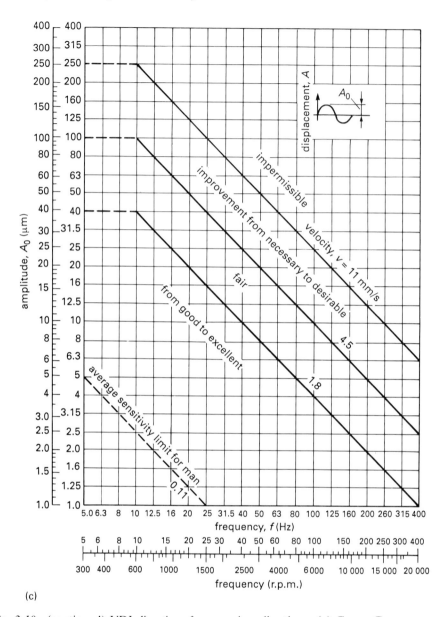

(c)

Fig. 2.10 (continued) VDI directives for assessing vibrations: (c) Group G.

The graphs and tables, which have been provided in this section, should enable the engineer to assess the likely effects of machine vibrations.

2.6 Dynamic effects due to blasting

Blasting operations in quarries or construction sites induce ground-borne

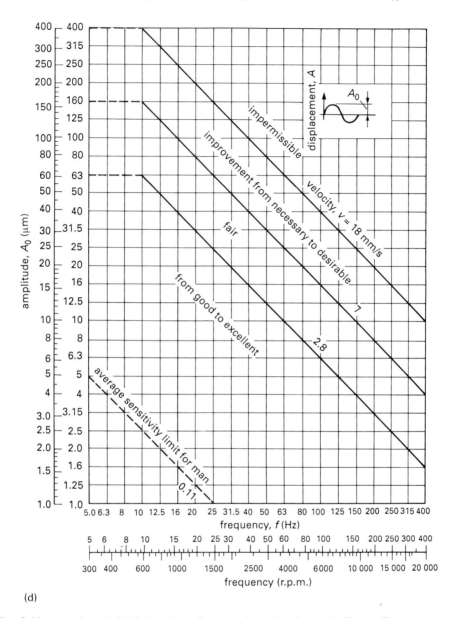

(d)

Fig. 2.10 (continued) VDI directives for assessing vibrations: (d) Group T.

vibrations. The associated dynamic loads and ground motion effects, although considerably smaller, are similar in form to those produced by an earthquake.

A considerable amount of information exists on structural damage which has been caused by blasting. In work published by the US Bureau of Mines (1936, 1937, 1938a, 1938b, 1940, 1942) the response of buildings and levels

Table 2.13 Permissible amplitudes for various machines (from Korchinski 1948).

	Harmless		Objectionable	
Machinery	Amplitude (mm)	Frequency per min.	Amplitude (mm)	Frequency per min.
(1) Weaving machines	0.30	140	1.00	—
(2) Spinning machines	0.10–0.12	200–550	—	—
(3) Confectionery			—	—
(chocolate press, biscuit	0.25	200		
press)	0.15	350–450		
(4) Forced draught and				
ventilating equipment	0.10	1 000	0.40	1 000
(5) Machine-tools				
(a) planer	0.35	—	—	—
(b) tapping machine	0.03			
milling machine				
slotting machine				
drilling machine				
grinding machine				
(6) Manufacture of ball-bearings				
(a) grinding machines	0.03	1 000		
(b) internal grinding	0.05			
(c) presses	0.09			
(d) inspection rooms with				
measuring instruments	0.005	2 000		

Table 2.14 Vibration severity ranges and examples of their application to small machines (Class I); medium-size machines (Class II); large machines (Class III); turbo machines (Class IV) (from BS 4675 Part 2 (1978).

Range classification	Ranges of vibration severity, RMS velocity (mm/s) at the range limits	Examples of quality			
		Class I	Class II	Class III	Class IV
0.28	0.28				
0.45	0.45	A			
0.71	0.71		A		
1.12	1.12	B		A	
1.8	1.8		B		A
2.8	2.8	C		B	
4.5	4.5		C		B
7.1	7.1	D		C	
11.2	11.2		D		C
18	18			D	
28	28				D
45	45				
71	71				

of vibration have been recorded. Rather than determining the magnitude of blasting loads on structures, certain allowable levels of vibrations are recommended by various authorities. These levels determine the maximum charge which can be fired safely at certain distances from structures situated on different soil conditions. It is often assumed in these situations that the amplitude of ground vibration will vary as the square root of the charge and inversely as the distance. Rules such as this have proven satisfactory under normal conditions. For example, the maximum charge to be fired at a distance of 1000 ft (305 m) to limit the maximum amplitude to 0.008 in. (0.2 mm) is 6400 lb (2900 kg).

Teichmann & Westwater (1957) have suggested a scale of permissible amplitudes to which various types of structure may be subjected during a blasting operation (Table 2.15).

An alternative approach to assess blasting effects has been suggested by Crandell (1949) who proposed that the kinetic energy be used as the criterion. The standard expression for the kinetic energy of a particular location may be arranged so that the only variable for a particular site is $\ddot{x}/f^2$, the energy ratio, $\ddot{x}$ being the acceleration and f the frequency. The caution stage is considered to have been reached when the energy ratio has a value of 3. Edwards & Northwood (1960), who have carried out controlled tests to deliberately damage buildings, agree in general with Crandell's proposals. They suggest a velocity criterion, however, as an alternative to the energy ratio. Edwards and Northwood consider that the threshold of damage corresponds to a velocity of 4.5 in./s (114 mm/s). The equivalent velocity which is based on Crandell's work is 3.3 in./s (84 mm/s). Table 2.16 gives the amplitudes of vibration which satisfy both criteria for a range of frequencies.

The effects of blasting may be minimized by delaying the charges so that explosive effects take place a few milliseconds apart. Details of blasting practice are to be found in *The Blasters' Handbook* (1969) and are also given by Langefors & Kiehlstrom (1963). Newmark & Hansen (1961) provide details for the design of blast-resistant structures.

Table 2.15 Maximum permissible amplitudes of blasting vibration (from Teichmann & Westwater 1957).

Classification	Description of property	Maximum amplitude (thousand of an inch) (mm in brackets)	
1	Structures of great value or frailty (e.g. ancient monuments, churches, and some badly designed properties)	4	(0.1)
2	Closely congregated houses, etc.	8	(0.2)
3	Isolated properties	16	(0.41)
4	Civil engineering structures	30[a]	(0.76)

[a] In very exceptional circumstances, e.g. for structures owned by the blasting company, an amplitude of 0.07 in. (1.78 mm) may be permitted.

Table 2.16 Limiting amplitudes for blasting vibration and damage.

Criterion	Amplitude (thousandths of an inch) at given frequency (cycles/second)					
	5	10	20	30	40	50
Crandell 'caution' ($\ddot{x}/f^2 = 3$)	105	53	26	18	13	11
Edwards–Northwood 'damage' (velocity = 4.5 in./s)	143	72	36	24	18	14

2.7 Dynamic effects due to road traffic, railway and piling sources

Structures incorporating methods of isolation to eliminate ground-borne vibration from underground railways have been constructed in London. In addition to these structures a section of underground railway track has been mounted upon special vibration isolators to reduce the vibration which would otherwise have been transmitted into adjacent buildings. Some details of these constructions are to be found in papers by Morton (1967), Waller (1966) and Grootenhuis (1967). The above are examples of the growing awareness of the necessity to eliminate ground-borne vibration from structures in industrial and urban areas where this type of vibration is objectionable.

Such a vibration is objectionable for two fundamental reasons. The first is with regard to the human aspect of tolerating conditions in which vibration is perceptible and this may be termed its 'nuisance value'. The Department of Energy Offshore Installations: *Guidance on Design, Construction and Certification* (HMSO 1990) recommends vibration limits for various categories of work and rest and is an example of vibration control based on human tolerance. Table 2.17 and Figs 2.11(a) and (b) provide information from the Department of Energy guidance notes.

The second reason concerns the economic aspect of the reduction in the value of the amenity. It is known, for instance, that the passage of road traffic or railway trains can interfere with sensitive instrumentation or impair the enjoyment of cinema performances, thus reducing the economic value of the affected structures.

Modern technological developments and constructional techniques have made the vibration isolation of structures a practical and economic possibility. The cost of such an operation has been shown to be of the same order as other services which make an adequate working and living environment.

The vibrations due to road, railway and piling sources have been measured both in structures and on the ground in the close vicinity of the sources. Although there have been recorded instances of structural damage to ancient

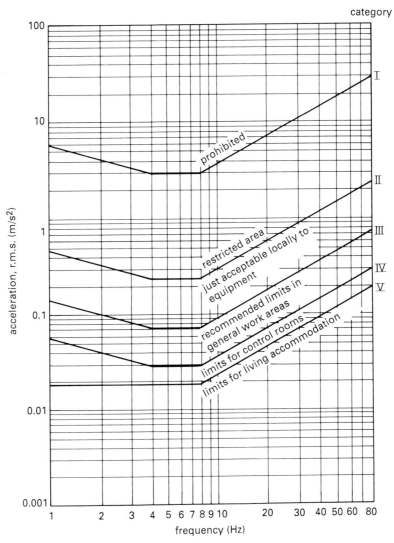

Fig. 2.11(a) Vibration limits for offshore installation 'vertical' axis (a_z). (From Department of Energy guidance notes, HMSO 1990.)

structures, particularly churches (Crockett 1963, Ciesielski 1963 and Rosivall & Goschy 1963) caused by the passage of road traffic, such ground-borne vibrations are not deemed to be damaging to modern structures. Similarly, the ground-borne vibration caused by railway trains may be considered to be a nuisance by the occupants of affected buildings but not necessarily damaging to the structure.

Badly controlled piling can cause structural damage particularly to building foundations which are adjacent to the operation (Alpan & Meidav 1963).

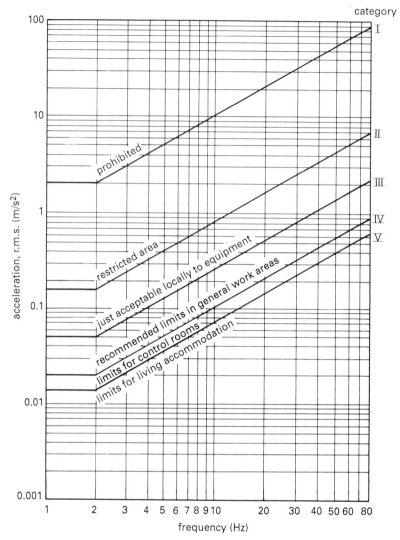

Fig. 2.11(b)   Vibration limits for offshore installations 'horizontal' axes (a_x, a_y). (From Department of Energy guidance notes, HMSO 1990.)

Since piling is a short-term operation, buildings are not especially designed to resist such effects. However, the engineer may be required to conduct a vibration survey of a structure to assess the potential damaging effects. The remaining part of this section provides an indication of the likely levels of vibration which may be associated with road traffic, railway and piling sources.

Road traffic vibrations result from the passage of vehicles over an irregular road surface. Measurements may be taken of the displacements, accelerations

Table 2.17   Descriptions of vibration limit categories (from Department of Energy guidance notes 1990).

Category	Description
I	Restricted area (less than 4 minutes' exposure) vibration limits. Short exposure to levels above these limits may create a health hazard and cause difficulty in walking. These high levels of vibration usually cause such alarm and discomfort that action is immediately and intuitively taken by persons subjected to the vibration. Vibration levels above these limits should be treated as prohibited.
II	Just acceptable locally to equipment, although vibration limits for machinery may be more restrictive than these levels. Annoyance and discomfort may be experienced.
III	Recommended design vibration limits for all general work areas. Vibration levels are easily detectable but not uncomfortable.
IV	Recommended design vibration limits for office, control rooms and similar areas.
V	Recommended design vibration limits for sleeping, recreation and similar accommodation areas. These vibration levels are just detectable

and frequencies caused by surface irregularities. Figure 2.12 shows typical acceleration–time vibrographs for the vertical component obtained from ground surface measurements; the vibration lasting for a few seconds as the vehicle passes the measurement position.

The traces obtained from railway vibrations are very similar, but are usually of greater magnitude and duration. The engineer requires to know the likely maximum values of displacement, acceleration and the associated frequencies which are transmitted to a particular location. Table 2.18 provides some typical examples of these values for both road traffic and railway vibrations. The information presented in this table may be compared with that provided by Steffens (1952) who has compiled similar information from a number of different sources.

The magnitude and frequencies of ground-borne vibrations which are caused by road traffic and railway sources are clearly a function of the vehicle suspension system, the surface roughness of the road or track and the properties of the soil through which the vibration is transmitted. It is possible, however, to extract some useful conclusions from the complex mechanism.

The frequencies which are generated by road traffic and railway vibrations are in the general range of 5–50 cycles/second. This range corresponds to typical values of the natural frequencies of structural components. There is some evidence to suggest that the frequencies transmitted through soils

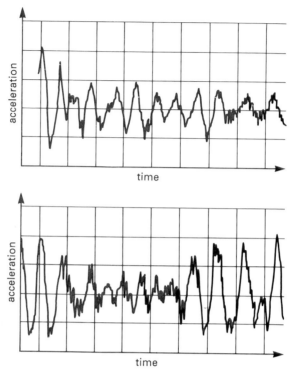

Fig. 2.12 Acceleration–time vibrographs of road traffic vibrations.

with high water contents (sands and silts) are lower than frequencies transmitted through more dense material (shale and rock). It would seem that only in extreme cases, which are close to the source, are vibrations likely to be classified into the 'perception' ranges on scales of assessment. Vibrations are generally required to exceed the 'painful' stage of assessment before structural damage is a possibility. According to the scales of intensity this situation requires occupants to leave buildings and it is obvious that road traffic and railway vibrations will not cause such effects in the short term. It has already been implied that the long-term effects of these vibrations

Table 2.18 Examples of road traffic and railway vibrations.

Authority	Source	Amplitude (inches)		Frequency (cycles/second)
		V	H	
BRS	Traffic	0.000 14	0.000 06	17–25
Mallock	⎱	0.001		10–15
BRS	⎰ Underground	0.20		10–30
Waller	⎰ railway	0.000 55		22

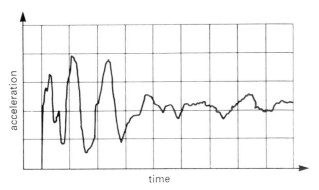

Fig. 2.13 Acceleration–time vibrograph of piling vibrations.

could, however, be a potential source of damage in the case of ancient structures.

Figure 2.13 shows an acceleration–time vibrograph for the vertical component of vibration caused by the impulsive blow of a piling hammer. These vibrations may be classified as 'clearly perceptible'. Vibrations which have been generated by the single blow of a piling hammer have few cycles of vibration and again exhibit the characteristic of producing lower transmitted frequencies through soils with high water contents. Once more, the transmitted frequencies are in the general range of the natural frequencies of structural components, i.e. 5–50 cycles/second. If damage is considered as a possibility as a result of pile driving, then continuous monitoring of affected structures is recommended together with a carefully planned piling programme.

2.7.1   The calculation of vibration magnitude at varying distances from the source

It is useful to be able to predict the value of say, the ground amplitude of vibration at some distance from a source using a limited amount of observed data. Although the transmission of ground-borne vibration is a complex phenomenon, an expression has been proposed (DEGEBO 1936, Barkan 1962) which may be used to predict the magnitude of ground-borne vibration from machinery, piling road traffic and railway sources. This expression is written as:

$$Z_{n+1} = Z_n \sqrt{\left(\frac{s_n}{s_{n+1}}\right)} \, e^{-\alpha(s_{n+1} - s_n)} \qquad (2.21)$$

where Z_{n+1}, Z_n are the magnitude of soil vibration at distances s_{n+1}, s_n from the source and α is the coefficient of wave energy absorption.

Table 2.19 Values of the coefficient of absorption (from Barkan 1962).

Soil	Coefficient of absorption
Yellow, water-saturated fine-grained sand	0.100
Yellow, water-saturated fine-grained sand in frozen state	0.060
Grey, water-saturated sand with laminae of peat and organic silt	0.040
Clayey sands with laminae of more clayey sands and of clays with some sand and silt, above groundwater level	0.040
Heavy water-saturated brown clays with some sand and silt	0.040–0.120
Marly chalk	0.100
Loess and loessial soil	0.100

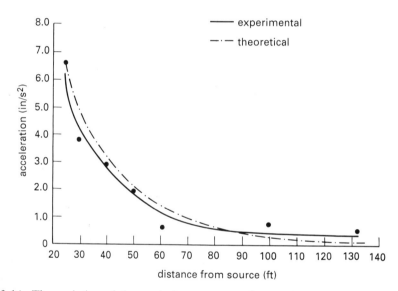

Fig. 2.14 The variation of the vertical component of ground acceleration due to railway vibrations.

For a perfectly elastic medium, $\alpha = 0$, and damping due to radiation only occurs. However, this is not the case for ground-borne vibrations and α has values depending upon the soil type to take account of internal damping. Table 2.19 gives the values provided by Barkan for some soil types.

Although Equation (2.21) is approximate, it may be used to predict the values of vibration at various distances from the source. Figure 2.14 shows the variation with distance of the vertical component of acceleration of the ground surface caused by railway vibrations. A value of 0.04 for the coefficient α was used in Equation (2.21).

Chapter Three
Systems With One Degree of Freedom

3.1 Introduction

The complexity of a dynamic system depends upon the number of degrees of freedom which are considered to be necessary to define the system's motion. SDOF systems represent, therefore, the simplest dynamic situation. Such a system, as discussed in Chapter 1, may be used to define the motion (Equation (1.2)). Equation (1.2) is a linear homogeneous differential equation of the second order. This equation may be considered to be the governing equation of the SDOF system and the form of the solution depends upon the mathematical representation for the forcing function $p(t)$.

Although the representation of a real structure by an idealized SDOF system, such as that discussed in Chapter 1, may be considered to be an oversimplification, useful results can be obtained, however, from the analysis of the simplified system. It is intended that some of the examples provided in this chapter adequately demonstrate how the behaviour of an apparently complex system may be described by a single co-ordinate.

The analysis of SDOF systems is provided by the solution to the differential equation of motion and this equation is obtained by applying one of the

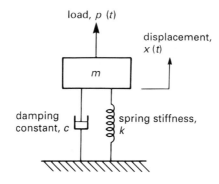

Fig. 3.1 Mass–spring–damper SDOF system.

47

methods discussed in Chapter 1. This chapter considers systems which are subject to either free vibration or to some type of forced vibration. The type of forced vibration depends upon the time variation of the forcing function and there are many types. It has been considered sufficient, however, to examine the response of SDOF systems which are subjected to harmonic, periodic, impulsive and general forcing functions.

The response of a system is governed by the amount of damping present. Consideration is given, therefore, to the types and amounts of damping in real structures and to the practical measurement of damping phenomena.

3.2 Free vibration response

The equation of motion for a SDOF mass–spring–damper system (Fig. 3.1) has been considered in Chapter 1 and may be written as

$$m\ddot{x}(t) + c\dot{x}(t) + kx(t) = p(t) \tag{3.1}$$

If the forcing function $p(t)$ of this equation equals zero, then the system defined by Equation (3.1) is subjected to free vibrations. The motion is caused by an initial displacement or velocity of the mass, i.e.

$$m\ddot{x}(t) + c\dot{x}(t) + kx(t) = 0 \tag{3.2}$$

The solution to this equation is a standard form and may be written as

$$x(t) = Ae^{bt} \tag{3.3}$$

Substitution into Equation (3.2) gives

$$(mb^2 + cb + k)\, Ae^{bt} = 0 \tag{3.4}$$

rearranging with $\omega^2 = k/m$

$$b^2 + \frac{c}{m}b + \omega^2 = 0 \tag{3.5}$$

therefore

$$b = -\frac{c}{2m} \pm \sqrt{\left[\left(\frac{c}{2m}\right)^2 - \omega^2\right]} \tag{3.6}$$

The value of b in Equation (3.6) and, therefore, $x(t)$ depends upon the damping constant, c. There are four alternative solutions to Equation (3.6) each representing a particular dynamic feature of the SDOF system.

3.2.1 Undamped motion

If the value of the damping constant c is equal to zero then the motion is undamped and the solution to Equation (3.6) is

$$b = \pm i\omega \tag{3.7}$$

The two alternative solutions to Equation (3.7) are included in the general solution of Equation (3.3), i.e.

$$x(t) = A_1 e^{i\omega t} + A_2 e^{-i\omega t} \tag{3.8}$$

since

$$e^{\pm i\omega t} = \cos \omega t \pm i \sin \omega t \tag{3.9}$$

then

$$x(t) = B_1 \sin \omega t + B_2 \cos \omega t \tag{3.10}$$

Equation (3.10) which represents simple harmonic motion may be re-written as

$$x(t) = B \sin (\omega t + \alpha) \tag{3.11}$$

The constants in Equations (3.10) and (3.11) may be obtained from the values of the displacement and velocity at zero time.

The quantity ω is the circular frequency or angular velocity and has units of radians per unit time. The frequency of vibration in cycles per second or hertz is given by

$$f = \frac{\omega}{2\pi} \tag{3.12}$$

and the natural period of vibration is

$$T = \frac{1}{f} = \frac{2\pi}{\omega} \tag{3.13}$$

The angle α in Equation (3.11) is the phase angle and the constant B is the maximum amplitude of motion. If the initial values of displacement $x(0)$ and velocity $\dot{x}(0)$ are substituted into Equation (3.10), then

$$B = (B_1^2 + B_2^2)^{1/2}$$

$$B = \left(\frac{\dot{x}(0)^2}{\omega^2} + x(0)^2 \right)^{1/2} \tag{3.14}$$

Figure 3.2 shows a graph of undamped free vibration.

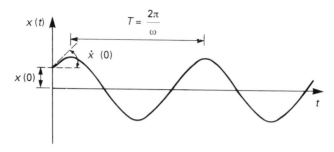

Fig. 3.2   Undamped free-vibration response.

Example 3.1

Show that for free undamped vibration the displacement $x(t)$ at any time t is given by $x(t) = B \sin(\omega_n t + \alpha)$, where B is the amplitude of the motion, ω_n is the natural circular frequency, and α is a phase angle.

A light beam of stiffness $k = 10^3$ N/m supports a mass of 4000 kg at its tip. The beam is undamped and set into free vibration. If the initial displacement is 5 mm and the displacement after 2 seconds is 4 mm, obtain the displacement after 3 seconds and the maximum amplitude of the motion.

To show that for free undamped vibration $x(t) = B \sin(\omega_n t + \alpha)$, the reader should establish the equation of motion for a SDOF system (Equation (3.1)) and the equation for b when the damping constant c is equal to zero (Equation (3.7)). Equation (3.11) which is the desired result follows directly from Equations (3.8), (3.9) and (3.10). Since

$$\omega^2 = \frac{k}{m}$$

$$\omega = \sqrt{\left(\frac{k}{m}\right)}$$

where k has units of N/m and m has units of kilograms, then

$$\omega = \sqrt{\left(\frac{1 \times 10^3}{4000}\right)} = 0.5 \text{ rad/s}$$

The displacement $x(0)$ and the velocity $\dot{x}(0)$ at time $t = 0$ are the initial conditions causing the free vibrations of the beam. It is seen that when the initial conditions are substituted into Equation (3.10), $B_1 = \dot{x}(0)/\omega$ and $B_2 = x(0)$, hence

$$x(t) = \frac{\dot{x}(0)}{\omega} \sin \omega t + x(0) \cos \omega t$$

At $t = 2$ seconds

$$4 = \frac{\dot{x}(0)}{0.5} \sin 0.5 \times 2 + 5 \cos 0.5 \times 2$$

$$\dot{x}(0) = 0.773 \text{ mm/s}$$

At $t = 3$ seconds

$$x(3) = \frac{0.773}{0.5} \sin 0.5 \times 3 + 5 \cos 0.5 \times 3$$

$$x(3) = 1.896 \text{ mm}$$

The maximum amplitude of the motion is given by

$$B = (B_1^2 + B_2^2)^{1/2} = \left[\left(\frac{0.773}{0.5} \right)^2 + 5^2 \right]^{1/2} = 5.23 \text{ mm}$$

Example 3.2

A mass M is attached to the midpoint of a light elastic cable 7.2 m long. The ends of the cable are attached to two points 8 m apart in the same horizontal plane. If the mass is subjected to a small displacement normal to the centreline of the cable, show that the resulting motion is simple harmonic and find its period. The modulus of elasticity of the cable is E and the area is A.

The tension in the cable when the mass M is displaced a small distance x (Fig. E3.1) is

$$F = \frac{EA}{3.6} \left[\sqrt{(16 + x^2)} - 3.6 \right]$$

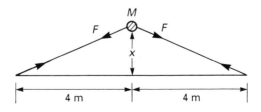

Fig. E3.1

The system is undamped and it is assumed that the mass of the cable is negligible. The elastic restoring force which is active on the displaced mass is

$$\frac{2Fx}{\sqrt{(16 + x^2)}} = \frac{2EAx}{3.6}\left[1 - \frac{3.6}{\sqrt{(16 + x^2)}}\right]$$

Since x is assumed to be small the x^2 under the square root sign may be neglected and the restoring force causing acceleration of the mass becomes $(EAx/18)$.

From D'Alembert's principle, the equation of motion for the mass is

$$M\ddot{x} + EAx/18 = 0$$

This is an equation of simple harmonic motion with period (Equation (3.13))

$$T = 2\pi\sqrt{\left(\frac{M}{k}\right)} = 26.66\sqrt{\left(\frac{M}{EA}\right)} \text{ s}$$

3.2.2 Damped motion

If the value of the damping constant c is non-zero then damping is present in the system and there are three alternative solutions for b in Equation (3.6) which correspond to the following conditions

$$\left(\frac{c}{2m}\right)^2 - \omega^2 = \begin{cases} < 0 & \text{(a)} \\ = 0 & \text{(b)} \\ > 0 & \text{(c)} \end{cases}$$ (3.15)

Case (a) occurs when $c < 2m\omega$ and corresponds to underdamped motion. For this case the roots of Equation (3.6) are complex and the solution of the equation of motion, Equation (3.2), is

$$x(t) = e^{(-ct/2m)}(C_1 \sin \omega_D t + C_2 \cos \omega_D t)$$ (3.16)

in which ω_D is the damped natural circular frequency. Since from case (b), $c = 2m\omega$, which may be redefined as c_c for the case of critical damping, the actual amount of damping in case (a) may be expressed in terms of a damping ratio ξ, where

$$\xi = \frac{c}{c_c} = \frac{c}{2m\omega}$$ (3.17)

and Equation (3.16) becomes

$$x(t) = e^{-\xi\omega t}(C_1 \sin \omega_D t + C_2 \cos \omega_D t)$$ (3.18)

or

$$x(t) = Ce^{-\xi\omega t}\sin(\omega_D t + \alpha) \tag{3.19}$$

The constants in Equation (3.19) are obtained by assuming the same initial conditions as before, hence

$$C = \left\{\left[\frac{\dot{x}(0) + x(0)\xi\omega}{\omega_D}\right]^2 + [x(0)]^2\right\}^{1/2} \tag{3.20}$$

and

$$\alpha = \tan^{-1}\left[\frac{\dot{x}(0) + x(0)\xi\omega}{\omega_D x(0)}\right] \tag{3.21}$$

The solution to the damped case may be thought of as a simple harmonic part which is multiplied by an exponential damping factor. Figure 3.3 illustrates the free vibration response of an underdamped system.

By considering the radical of the equation for b, an expression for the damped vibration frequency is obtained, i.e.

$$b = -\xi\omega \pm \sqrt{[(\xi\omega)^2 - \omega^2]} \tag{3.22}$$

which may be written as

$$b = -\xi\omega \pm i\omega_D \tag{3.23}$$

where

$$\omega_D = \omega\sqrt{(1 - \xi^2)} \tag{3.24}$$

In most practical structures in which ξ is small the damped and undamped frequencies are considered to be identical.

Consider further Fig. 3.3 from which it can be seen that the amplitudes

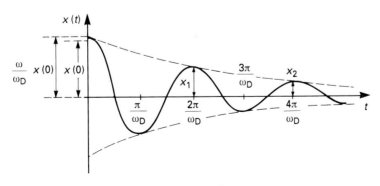

Fig. 3.3   Free vibration response of underdamped system.

x_1, x_2, ... x_n, etc., occur in successive time intervals T_D. By combining Equations (3.16) and (3.17), the ratio of successive peaks is

$$\frac{x_n}{x_{n+1}} = \frac{e^{-ct/2m}}{e^{-c(t+T_D)/2m}} = e^{2\pi\xi(\omega/\omega_D)} \tag{3.25}$$

The natural logarithm of this ratio is called the logarithmic decrement δ, i.e.

$$\delta = \ln\left(\frac{x_n}{x_{n+1}}\right) = 2\pi\xi\left(\frac{\omega}{\omega_D}\right) \tag{3.26}$$

Substituting Equation (3.24) in Equation (3.26) gives

$$\delta = \frac{2\pi\xi}{\sqrt{(1 - \xi^2)}} \tag{3.27}$$

Equation (3.27) is approximately equal to $2\pi\xi$ for small amounts of damping and in this instance Equation (3.25) may be expressed as a series in which only the first two terms are retained, i.e.

$$\frac{x_n}{x_{n+1}} = e^{2\pi\xi} = 1 + 2\pi\xi \tag{3.28}$$

and the damping ratio is obtained from

$$\xi = \frac{x_n - x_{n+1}}{2\pi x_{n+1}} \tag{3.29}$$

In lightly damped systems it is often more accurate to consider the amplitude response of peaks which are several cycles apart, say m cycles; the damping ratio is then obtained from

$$\ln\left(\frac{x_n}{x_{n+m}}\right) = 2\pi m\xi\left(\frac{\omega}{\omega_D}\right) \tag{3.30}$$

which may be approximated to the following equation for lightly damped systems

$$\xi = \frac{x_n - x_{n+m}}{2\pi m x_{n+m}} \tag{3.31}$$

Case (b) occurs as has been stated when $c_c = 2m\omega$. The system is said to be critically damped. For this condition Equation (3.6) becomes

$$b = \frac{-c}{2m} = -\omega \tag{3.32}$$

and the solution to the equation of motion is

$$x(t) = e^{-\omega t}(D_1 + D_2 t) \tag{3.33}$$

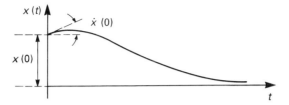

Fig. 3.4   Free vibration response with critical damping.

The constants are once again determined from the initial conditions. A plot of Equation (3.33) is shown in Fig. 3.4. It should be noted that there is no oscillation of the system about the mean position. Upon displacement, the mass will return directly to its static position.

Case (c) is considered here for completeness although the corresponding overdamped system is unlikely to be of practical interest to structural engineers. For this case the two solutions of Equation (3.6) are real and may be written as

$$b = -\xi\omega \pm \bar{\omega}_D \tag{3.34}$$

in which

$$\bar{\omega}_D = \omega\sqrt{(\xi^2 - 1)} \qquad \text{where } \xi > 1$$

The solution of the equation of motion becomes

$$x(t) = e^{-\xi\omega t}\,(E_1 \sinh \bar{\omega}_D t + E_2 \cosh \bar{\omega}_D t) \tag{3.35}$$

where the constants are evaluated from the initial conditions and the motion is not oscillatory.

Example 3.3

The beam BD of the portal frame shown in Fig. E3.2 is subjected to a horizontal displacement of 5 mm and then suddenly released. At the end of the first cycle of motion the horizontal displacement of the beam was found to be 4 mm. If the modulus of elasticity of the material of the frame is 200 kN/mm^2, the second moment of area of the columns is 6×10^8 mm^4 and the mass of the beam is 4000 kg, calculate:

(a)   the natural frequency of the frame assuming that there is no damping;
(b)   the ratio of the damped and undamped periods of motion.

The translational stiffness of the members AB and CD is given by $12EI/l^2$. Hence

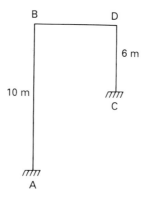

Fig. E3.2

$$k_{AB} = \frac{12 \times 200 \times 6 \times 10^8}{(10 \times 10^3)^3} \text{kN/mm} = 1.44 \times 10^6 \text{N/m}$$

Similarly

$$k_{CD} = \frac{12 \times 200 \times 6 \times 10^8}{(6 \times 10^3)^3} \text{kN/mm} = 6.67 \times 10^6 \text{N/m}$$

The translational stiffness of the complete frame is given by

$$k_{AB} + k_{CD} = 8.11 \times 10^6 \text{N/m}$$

The natural circular frequency is obtained from

$$\omega = \sqrt{(k/m)} = \sqrt{(8.11 \times 10^6/4\,000)} = 45 \text{ rad/s}$$

and the undamped natural frequency of the frame is obtained from Equation (3.12)

$$f = \omega/2\pi = 45/2\pi = 7.16 \text{ hertz}$$

From Equation (3.24), the ratio of the damped and undamped circular frequencies is given by

$$\omega_D/\omega = \sqrt{(1 - \xi^2)}$$

therefore, the ratio of the damped and undamped periods is

$$T_D/T = 1/\sqrt{(1 - \xi^2)}$$

where $\xi = \delta/2\pi$ for small amounts of damping (Equation (3.27)) and the logarithmic decrement $\delta = \ln(5/4) = 0.223$ (Equation (3.26)). Thus the damping ratio $\xi = 0.223/2\pi = 3.55\%$ and

$$T_D/T = 1/\sqrt{(1 - 0.355^2)} = 1.000\,65$$

3.3 Forced vibration response

If it is required to analyze a system during the application of a vibration source, the forcing function, $p(t)$, in the equation of motion is not zero and must be represented by a suitable mathematical expression.

3.3.1 Response to harmonic loading

Many sources of vibration may be represented by harmonic functions. One such function allows $p(t)$ to be represented by $p_0 \cos \dot{\omega}t$ (where $\dot{\omega}$ is the forcing frequency) and the corresponding equation of motion becomes

$$m\ddot{x}(t) + c\dot{x}(t) + kx(t) = p_0 \cos \dot{\omega}t \tag{3.36}$$

The solution to this equation is the sum of two parts; the complementary function $x_c(t)$, and the particular integral $x_p(t)$. The complementary function is obtained from the solution of the equation of motion when the forcing function is zero and, therefore, corresponds to the free vibration solution of the previous section. The particular integral is the direct solution without constants of Equation (3.36) and represents the specific behaviour of the system which is caused by the dynamic load.

Therefore,

$$x(t) = x_c(t) + x_p(t) \tag{3.37}$$

It was stated in the previous section that the free vibration solution depends upon the amount of damping present, but in practice structural engineers are usually concerned with underdamped systems. For this case the complementary solution is given by Equation (3.18) which represents an exponentially damped transient response, i.e.

$$x_c(t) = e^{-\xi \omega t}(C_1 \sin \omega_D t + C_2 \cos \omega_D t) \tag{3.38}$$

To obtain the particular integral it is assumed that $x_p(t) = C_3 \cos \dot{\omega}t + C_4 \sin \dot{\omega}t$. This equation is substituted into Equation (3.36) and the coefficients of $\cos \dot{\omega}t$ and $\sin \dot{\omega}t$ are equated separately to give

$$-m\dot{\omega}^2 C_3 + c\dot{\omega}C_4 + kC_3 = p_0$$

$$-m\dot{\omega}^2 C_4 - c\dot{\omega}C_3 + kC_4 = 0$$

Eliminating C_3 and C_4 and solving eventually gives the steady-state response

$$x_p(t) = \frac{p_0 \cos (\dot{\omega}t - \alpha)}{[(k - m\dot{\omega}^2)^2 + c^2\dot{\omega}^2]^{1/2}} \tag{3.39}$$

and

$$\tan \alpha = \frac{c\dot{\omega}}{k - m\dot{\omega}^2} \tag{3.40}$$

On substituting $\xi = c/(2m\omega)$, $\omega^2 = k/m$ and $\beta = \dot{\omega}/\omega$, the ratio of the forced vibration frequency to the undamped frequency of the system, Equations (3.39) and (3.40) become

$$x_p(t) = \frac{p_0}{k} \frac{\cos(\dot{\omega}t - \alpha)}{[(1 - \beta^2)^2 + (2\xi\beta)^2]^{1/2}} \tag{3.41}$$

$$\alpha = \tan^{-1} \frac{2\xi\beta}{1 - \beta^2} \tag{3.42}$$

The complete solution to the equation of motion is, then,

$$x(t) = e^{-\xi\omega t}(C_1 \sin \omega_D t + C_2 \cos \omega_D t) + \frac{p_0}{k} \frac{\cos(\dot{\omega}t - \alpha)}{[(1 - \beta^2)^2 + (2\xi\beta)^2]^{1/2}} \tag{3.43}$$

The constants C_1 and C_2 are obtained as before from the initial conditions, however the transient part of the solution is damped out within a few cycles and is usually of little interest. It is the steady-state response at the frequency of the applied load $\dot{\omega}$ which is important and only this part of the response will be considered.

$$x(t) = \frac{p_0}{k} \frac{\cos(\dot{\omega}t - \alpha)}{[(1 - \beta^2)^2 + (2\xi\beta)^2]^{1/2}} \tag{3.44}$$

From Equation (3.41) it will be noted that p_0/k equals the static deflection x_s of the system subjected to force p_0. Thus

$$\frac{x(t)}{x_s} = \frac{\cos(\dot{\omega}t - \alpha)}{[(1 - \beta^2)^2 + (2\xi\beta)^2]^{1/2}} \tag{3.45a}$$

The maximum amplitude is when $\cos(\dot{\omega}t - \alpha) = 1$, i.e.

$$\frac{x(t)}{x_s} = \frac{1}{[(1 - \beta^2)^2 + (2\xi\beta)^2]^{1/2}} \tag{3.45b}$$

and is called the dynamic magnification factor. The angle α is the phase angle and is the amount by which the response lags behind the disturbing force.

It is instructive to plot Equations (3.42) and (3.45) and these have been shown in Figs 3.5 and 3.6. Figure 3.5 shows the dynamic magnification factor plotted against the frequency ratio for various values of the damping ratio. A number of interesting points may be observed from this figure.

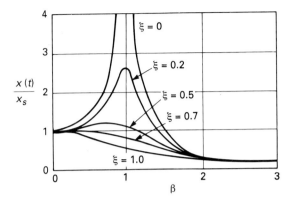

Fig. 3.5 Variation of dynamic magnification factor with damping and frequency.

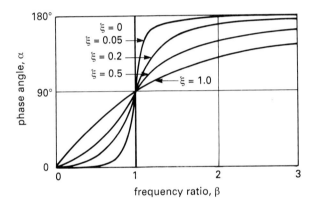

Fig. 3.6 Variation of phase angle with damping and frequency.

There are limiting values of unity and zero for the dynamic magnification factor which correspond to the low and high values respectively of the frequency ratio for all values of damping ratio. For a particular value of the damping ratio, the maximum value of the dynamic magnification factor corresponds to a frequency ratio which is very close to unity. The condition of resonance occurs when the forcing frequency and the natural frequency coincide. This condition should obviously be avoided for lightly damped systems. The subject of damping will be considered later in this chapter in more detail.

Figure 3.6 shows how the phase angle varies with the frequency ratio and damping ratio.

The right-hand side of Equation (3.36) may be replaced by $p_0 \sin \omega t$, in which case the complementary function remains the same and the particular integral is changed. The particular integral is obtained by replacing the $\cos(\omega t - \alpha)$ term in Equation (3.43) by $\sin(\omega t - \alpha)$.

If the motion given by Equation (3.36) is considered to be undamped then the particular integral is obtained by assuming

$$x_p(t) = E \cos \dot{\omega} t \qquad (3.46)$$

The amplitude E is obtained by substituting Equation (3.46) into the equation of motion

$$m\ddot{x}(t) + kx(t) = p_0 \cos \dot{\omega} t \qquad (3.47)$$

to give

$$-m\dot{\omega}^2 E \cos \dot{\omega} t + kE \cos \dot{\omega} t = p_0 \cos \dot{\omega} t \qquad (3.48)$$

from which

$$E = \frac{p_0}{k(1 - \beta^2)} \qquad (3.49)$$

and the particular integral solution becomes

$$x_p(t) = \frac{p_0}{k} \frac{\cos \dot{\omega} t}{(1 - \beta^2)} \qquad (3.50)$$

If the forcing function had been given by $p_0 \sin \dot{\omega} t$ then the particular integral would have been given by

$$x_p(t) = \frac{p_0}{k} \frac{\sin \dot{\omega} t}{(1 - \beta^2)} \qquad (3.51)$$

The complete solution to the undamped equation of motion is obtained by adding either Equation (3.50) or (3.51) to the complementary function given by Equation (3.10).

Example 3.4

A single degree of freedom portal frame is subjected to a disturbing force of $0.01\omega^2 \sin \omega t$. If the effective mass of the frame participating in the motion is $1\,000\,kg$, the viscous damping coefficient is $5.0 \times 10^2\,N/m/s$ and the spring stiffness is $2.5 \times 10^4\,N/m$. Determine the value of the disturbing force frequency to give the maximum amplitude of steady forced vibration and the value of this amplitude.

The form of the disturbing force which is given in this example is typical of the out of balance component due to a rotating machine and replaces the right-hand side of Equation (3.36). The solution of this equation enables the response of the portal frame to be obtained. The details are left to the reader, but the procedure follows that adopted from Equation (3.36) to Equation (3.44). For a general disturbing force of $F\omega^2 \sin \omega t$ acting on a mass m, the amplitude X of the steady-state forced vibration may be shown to be

$$X = \frac{F}{m} \frac{\beta^2}{[(1 - \beta^2)^2 + (2\xi\beta)^2]^{1/2}}$$

with the usual notation. For this case the amplitude is related to the displacement as follows

$$x(t) = X \sin(\omega t - \alpha)$$

In order to determine the value of ω to give a maximum amplitude (in this case a maximum value of mX/F), $d(X_n/F)/d\beta = 0$ must be evaluated. The frequency is obtained after some manipulation and is obtained from

$$\beta = \omega/\omega_n = (1 - 2\xi^2)^{-1/2}$$

Now

$$\omega_n = \sqrt{(k/m)} = \sqrt{(2.5 \times 10^4/10^3)} = 5 \text{ rad/s}$$

The critical damping coefficient

$$c_c = 2m\omega_n = 2 \times 1000 \times 5 = 1.0 \times 10^4 \text{ N/m/s}$$

The damping ratio

$$\xi = c/c_c = 5 \times 10^2/1 \times 10^4 = 0.05$$

Therefore

$$\omega = \omega_n(1 - 2\xi^2)^{-1/2} = 5(1 - 2 \times 0.05^2)^{-1/2} = 5.013 \text{ c/s}$$

The corresponding amplitude is

$$X = \frac{0.01}{1000} \frac{1.0025^2}{[(1 - 1.0025^2)^2 + (2 \times 0.05 \times 1.0025^2)]^{1/2}} = 0.1 \text{ mm}$$

Example 3.5

Figure E3.3 shows a SDOF system. If the base is subjected to a motion of $X \sin \bar{\omega} t$ investigate the motion of the system and discuss its application to the measurement of vibration.

This problem is of particular interest since its solution forms the basis of the theory of vibration measurement for seismic instrumentation.

If the relative displacement between the mass and its base is x, then

$$x(t) = x_1(t) - x_2(t)$$

and since $x_2(t) = X \sin \bar{\omega} t$, the equation of motion becomes

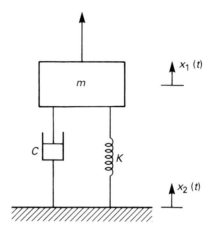

Fig. E3.3

$$m\ddot{x}(t) + c\dot{x}(t) + kx(t) = mX\dot{\omega}^2\sin\omega t$$

The steady-state solution for this equation may be obtained by suitably modifying Equation (3.44), i.e.

$$x(t) = X\frac{\beta^2\sin(\dot{\omega}t - \alpha)}{[(1 - \beta^2)^2 + (2\xi\beta)^2]^{1/2}}$$

and α is given by Equation (3.42). The amplitude of the relative motion is

$$X\beta^2/[(1 - \beta^2)^2 + (2\xi\beta)^2]^{1/2}$$

The system may be used to measure amplitude of vibration for which the natural frequency of the associated instrument is low and β is therefore large. For this case the amplitude of the relative motion is X and the phase angle α is $\tan^{-1}0$ or $180°$, which means that the displacements of the mass and the base are $180°$ out of phase.

The system may also be used to measure acceleration for which the natural frequency is high. The amplitude of relative motion now becomes equal to $X\beta^2$ or $X\dot{\omega}^2/\omega_n$. Since the acceleration of the base is $X\dot{\omega}^2$, it can be seen that the amplitude of the relative motion gives a value for the acceleration.

Further consideration of vibration instrumentation and measurement will be dealt with in Chapter 6.

3.4 Periodic vibration response

Any arbitrary complex periodic loading system such as that shown in Fig. 1.1 may be expressed as a Fourier series.

Each term in the series represents a harmonic loading component and the response to the total loading is the sum of the responses of each harmonic component. The expressions for the response of a SDOF system to individual harmonic loads have been obtained in the previous section and the same expressions will be used to obtain the response to periodic loading.

The Fourier series for an arbitrary periodic loading may be expressed by

$$p(t) = a_0 + \sum_{n=1}^{\infty} a_n \cos \omega_p nt + \sum_{n=1}^{\infty} b_n \sin \omega_p nt \tag{3.52}$$

where $\omega_p = 2\pi/T_p$ and T_p is the period of the load function.

The coefficients in the series are given by

$$a_0 = \frac{1}{T_p} \int_0^{T_p} p(t)\, dt \tag{3.53}$$

$$a_n = \frac{2}{T_p} \int_0^{T_p} p(t) \cos \omega_p nt\, dt \tag{3.54}$$

$$b_n = \frac{2}{T_p} \int_0^{T_p} p(t) \sin \omega_p nt\, dt \tag{3.55}$$

When considering an undamped SDOF system it should be remembered that the complementary solution is unaffected by the following argument, but must be added to the result to obtain the total response. The steady-state response of a SDOF undamped system to the sine and cosine terms in a harmonic loading series is obtained from Equations (3.50) and (3.51).

For each sine term the steady-state response is given by

$$x_n(t) = \frac{b_n}{k} \frac{\sin n\dot{\omega}_1 t}{(1 - \beta_n^2)} \tag{3.56}$$

where $\beta_n = n\dot{\omega}_1/\omega = \dot{\omega}_n/\omega$ is the frequency of the nth harmonic load component. Similarly, for each cosine term the steady-state response is given by

$$x_n(t) = \frac{a_n}{k} \frac{\cos n\dot{\omega}_1 t}{(1 - \beta_n^2)} \tag{3.57}$$

The response to the constant part of the load series is given by the static deflection, i.e.

$$x_0 = \frac{a_0}{k} \tag{3.58}$$

The total response of the system is obtained by summing the individual response components, i.e.

$$x(t) = \frac{1}{k} \left[a_0 + \sum_{n=1}^{\infty} \frac{1}{(1 - \beta_n^2)} (a_n \cos n\dot{\omega}_1 t + b_n \sin n\dot{\omega}_1 t) \right] \tag{3.59}$$

The complementary function must be added to Equation (3.59). For a damped SDOF system it is only necessary to consider the steady-state response. For this case, the relevant expressions are given by Equation (3.44) and the corresponding equation with $\cos(\dot{\omega}t - \alpha)$ replaced by $\sin(\dot{\omega}t - \alpha)$. These equations may be written as

$$x_n(t) = \frac{b_n}{k} \frac{1}{(1 - \beta_n^2)^2 + (2\xi\beta_n)^2} \left[(1 - \beta_n^2)\sin n\dot{\omega}_1 t - 2\xi\beta_n \cos n\dot{\omega}_1 t \right]$$

(3.60)

for each harnomic sine term, and

$$x_n(t) = \frac{a_n}{k} \frac{1}{(1 - \beta_n^2)^2 + (2\xi\beta_n)^2} \left[(1 - \beta_n^2)\cos n\dot{\omega}_1 t + 2\xi\beta_n \sin n\dot{\omega}_1 t \right]$$

(3.61)

for each harmonic cosine term.

Therefore, the total response of the system is given by

$$x(t) = \frac{1}{k} \left\{ a_0 + \sum_{n=1}^{\infty} \frac{1}{(1 - \beta_n^2)^2 + (2\xi\beta_n)^2} \left[(a_n 2\xi\beta_n + b_n(1 - \beta_n^2)) \right. \right.$$

$$\left. \left. \sin n\dot{\omega}_1 t + (a_n(1 - \beta_n^2) - b_n 2\xi\beta_n)\cos n\dot{\omega}_1 t \right] \right\}$$

(3.62)

3.5 Impulsive load response

Impulsive or shock loading provides a further class of dynamic loading which is characterized by a very short time history. Since the maximum response of a structure is attained very quickly the damping forces present do not have sufficient time to resist the motion. For this reason only undamped motion will be considered in this section.

Although actual shock loading may be complex, the response of a SDOF system to simpler idealized shock motions produces results which are of engineering significance. This is particularly so when the results are presented in the form of the shock spectra which are discussed in section 3.6. The discrete impulse and the step impulse are the limiting cases of shock motion; the half-sine wave, decaying sine wave and the triangular impulse are other forms of acceptable shock idealization.

3.5.1 The impulse or rectangular step

Figure 3.7 shows a step impulse which may be called a discrete impulse if its duration is very short. There are two distinct phases to be considered

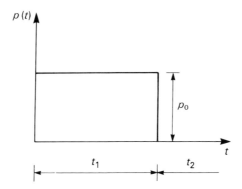

Fig. 3.7   Rectangular step impulse.

during analysis and these correspond to the period during which the pulse acts upon the system and the subsequent free vibrations. These phases correspond to the times t_1 and t_2 which are shown in Fig. 3.7.

The engineer is more interested in the maximum response to impulsive or shock loads than a complete time history of response. The maximum response can occur during the time of the pulse or during the subsequent free vibrations. The following analysis demonstrates that the ratio of the pulse duration to the system natural frequency determines when the maximum response occurs.

Consider the case when $t < t_1$; the impulsive force is constant and equals p_0. The undamped equation of motion is

$$m\ddot{x}(t) + kx(t) = p_0 \tag{3.63}$$

It may be shown that the particular solution is given by

$$x_p(t) = \frac{p_0}{k} \tag{3.64}$$

and the general solution is

$$x(t) = B_1 \sin \omega t + B_2 \cos \omega t + \frac{p_0}{k} \tag{3.65}$$

If it is assumed that the system was originally stationary, then $x(0) = \dot{x}(0) = 0$ and $B_1 = 0$, while $B_2 = p_0/k$. Equation (3.65) becomes

$$x(t) = \frac{p_0}{k} (1 - \cos \omega t) \tag{3.66}$$

and represents the response of the system for $t \leq t_1$. If Equation (3.66) is differentiated and equated to zero, it will be found that $t = \pi/\omega$ and the maximum response is

$$x_{\max}(t) = \frac{2p_0}{k} \tag{3.67}$$

Since p_0/k represents static deflection, x_s, the dynamic magnification factor is 2. It may be inferred that the maximum response to the rectangular impulse will always occur during its application provided that $t_1 \geq T/2$, where T, the natural period of the system, equals $2\pi/\omega$.

When $t > t_1$, the pulse has ceased and the free vibration condition exists with the response $x(t)$ given by either Equation (3.10) or Equation (3.11). The maximum value of the response is given by

$$x_{\max}(t) = B = \sqrt{(B_1^2 + B_2^2)} \tag{3.68}$$

If a new variable $t = (t_2 - t_1)$ is introduced, then the constants are $B_1 = \omega(p_0/k)\sin \omega t_1$, and $B_2 = (p_0/k)(1 - \cos \omega t_1)$. By substituting these values into Equation (3.68) and after some manipulation this becomes

$$x_{\max}(t) = \frac{p_0}{k} 2\sin\left(\frac{\pi t_1}{T}\right) \tag{3.69}$$

This equation is valid so long as the duration of the pulse is not greater than half the natural period of the system, i.e. $t_1 \leq T/2$.

3.5.2 Half-sine wave impulse

Figure 3.8 shows a half-sine wave impulse $p(t) = p_0 \sin \omega t$.

During the time $t < t_1$, the pulse is acting on the SDOF structure and subjecting it to forced harmonic motion. The response of the system during this time is obtained by combining Equations (3.10) and (3.51). If the initial conditions of displacement and velocity are assumed to be zero, then the undamped response becomes

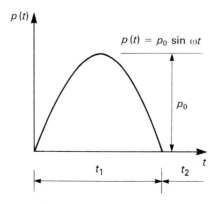

Fig. 3.8 Half-sine wave impulse.

$$x(t) = \frac{p_0}{k} \frac{1}{(1 - \beta^2)} (\sin \dot{\omega} t - \beta \sin \omega t) \tag{3.70}$$

Upon completion of the pulse the system undergoes free vibrations during $t_2 = (t - t_1)$. The initial conditions of displacement and velocity for the free vibrations correspond to those existing at the end of the pulse, i.e. $x(t_1)$ and $\dot{x}(t_1)$. Equation (3.10) is the undamped free vibration response, i.e.

$$x(t_2) = \frac{\dot{x}(t_1)}{\omega} \sin \omega t_2 + x(t_1) \cos \omega t_2 \tag{3.71}$$

Differentiating Equation (3.70) to determine the time of occurrence of the maximum response yields

$$\cos \dot{\omega} t = \cos \omega t \tag{3.72}$$

or

$$\dot{\omega} t = 2\pi n \pm \omega t \qquad n = \pm 0, 1, 2, \ldots \tag{3.73}$$

Equations (3.72) and (3.73) are valid for $\dot{\omega} t \le \pi$ for which the maximum response occurs during the impulse. Equation (3.73) becomes

$$\dot{\omega} t = 2\pi \Big/ \left(1 + \frac{1}{\beta}\right) \tag{3.74}$$

Equation (3.74) is substituted into Equation (3.70) to obtain the maximum response. If $\beta > 1$, the maximum response occurs during free vibration of the system. Since $\dot{\omega} t = \pi$, the initial conditions for the free vibrations become

$$x(t_1) = -\frac{p_0}{k} \frac{1}{(1 - \beta^2)} \beta \sin \pi/\beta \tag{3.75}$$

$$\dot{x}(t_1) = -\frac{p_0}{k} \frac{\dot{\omega}}{(1 - \beta^2)} (1 + \cos \pi/\beta) \tag{3.76}$$

and the amplitude of the free vibration is given by Equation (3.68), which becomes in this case

$$x_{max}(t) = \frac{p_0}{k} \frac{1}{(1 - \beta^2)} 2\beta \cos \pi/2\beta \tag{3.77}$$

3.6 Shock or response spectra

The maximum undamped response of a SDOF system which is subjected to a shock motion depends upon the ratio of the impulse duration to the natural period of vibration of the system, i.e. t_1/T. The equations for the maximum response, which have been given above, may be divided by the static deflection p_0/k, to give an expression for the dynamic magnification

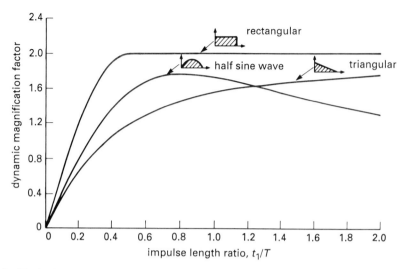

Fig. 3.9 Typical shock spectra.

factor $x_{max}(t)/x_s$. The engineer finds it particularly convenient to plot the dynamic magnification factor against values of t_1/T for various forms of shock motion. Such plots are known as shock or response spectra and they predict the maximum response of a system to a given shock motion. Figure 3.9 shows the spectra for three types of impulsive motion (the rectangular and half-sine wave impulses have been considered in the previous section).

Section 3.5 assumes that the shock motions are applied to the mass, however, alternative idealizations may be preferable for shock analysis which requires the motion to be applied to the base of the system.

Let it be assumed that the SDOF system in Fig. 3.1 is subjected to a base displacement of x_b and x is the relative displacement of the system with respect to the base. For this case the equation of motion is

$$m\ddot{x} + c\dot{x} + kx = -m\ddot{x}_b \tag{3.78}$$

The effect of the base motion is, therefore, equivalent to subjecting the system to a force which is given by the product of the mass and base acceleration.

For maximum impulsive conditions the equivalent static displacement is $m\ddot{x}_{bmax}/k$ and the dynamic magnification factor after ignoring the negative sign for convenience becomes

$$\frac{x_{max}}{m\ddot{x}_{bmax}/k} \qquad \text{or} \qquad \frac{\ddot{x}_{max}}{\ddot{x}_{bmax}}$$

where $\ddot{x}_{max}$ is the total mass acceleration in an undamped system.

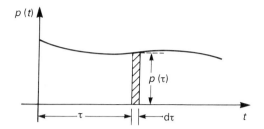

Fig. 3.10   General loading for the derivation of the Duhamel integral.

The maximum total acceleration of the mass $\ddot{x}_{max}$ is given by the product of the spring stiffness k and the total mass displacement x_{max}. Thus graphs of shock spectra, of which those in Fig. 3.9 are typical, may be used to assess the response of idealized systems to impulses which may be applied to either the mass or the base.

3.7 General load response

This chapter has considered the response of SDOF systems which have been subjected to forcing functions represented by particular mathematical expressions. In many cases these expressions provide a sufficiently accurate description of a forcing function. There are, however, occasions when such exact expressions cannot be used and an alternative approach becomes necessary. The description of a general loading $p(t)$, such as that shown in Fig. 3.10, is not conveniently represented by an exact expression.

3.7.1 Undamped motion

The procedure to obtain the response of an undamped SDOF system, which is initially at rest, will now be considered. If the general loading defined by $p(t)$ acts on the mass for an increment of time δt, the corresponding impulse is $p(t)\delta t$. The corresponding change in momentum of the mass is $m\dot{x}(t)$ where the term $kx(t)\,\delta t$ in the impulse momentum relationship is assumed to be negligibly small as δt approaches zero. Hence

$$p(t)\,\delta t = m\dot{x}(t) \tag{3.79}$$

For $t > 0$, free vibrations will occur and the response will be given by Equation (3.10). With initial conditions $x(0) = 0$ and $\dot{x}(0) = p(t)\,\delta t/m$, Equation (3.10) becomes

$$\Delta x(t) = \frac{p(t)\,\delta t}{m\omega}\sin\omega t$$

If the impulse had occurred at $t=\tau$ and lasted for a duration $\delta\tau$ (Fig. 3.10), the response at some time t_1, where $t_1 = (t - \tau)$, is given by

$$\Delta x(t_1) = \frac{p(\tau)\,\delta\tau}{m\omega} \sin \omega(t - \tau) \tag{3.80}$$

The total response is given by the summation of all such impulses, i.e.

$$x(t_1) = \frac{1}{m\omega} \int_0^t p(\tau) \sin \omega(t - \tau)\, d\tau \tag{3.81}$$

Equation (3.81) is called the Duhamel integral and assumes that the initial displacement and velocity of the mass are zero at $t=0$. If different initial conditions exist, the corresponding free vibration solution must be added to Equation (3.81).

In a general loading situation, Equation (3.81) is unlikely to be integrated explicitly and will require evaluation using a numerical integration rule. Using the trigonometric identity $\sin(\omega t - \omega\tau) = \sin \omega t \cos \omega\tau - \cos \omega t \sin \omega\tau$, Equation (3.81) becomes

$$x(t) = \frac{\sin \omega t}{m\omega} \int_0^t p(\tau) \cos \omega\tau\, d\tau - \frac{\cos \omega t}{m\omega} \int_0^t p(\tau) \sin \omega\tau\, d\tau \tag{3.82}$$

in which the integrals may be evaluated numerically using Simpson's rule. Other integration schemes may be used, but Simpson's rule provides acceptable results without excessive calculation.

Simpson's rule states that if the range of integration $(b - a)$ for a function $y(\tau)$ is divided into an even number, n, of equal parts each of width $(b - a)/n$, called the interval $\Delta\tau$, and ordinates are erected at $\tau = a$ and $\tau = b$ and at each point of the subdivision, then

$$\int_a^b y(\tau)\, d\tau \simeq \frac{1}{3} \times \text{interval} \times (\text{sum of first and last ordinates}$$

$$+ \text{ four times sum of odd ordinates}$$
$$+ \text{ twice sum of even ordinates})$$

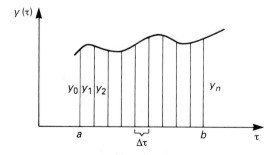

Fig. 3.11   Arrangement for Simpson's integration.

From Fig. 3.11,

$$\int_a^b y(\tau)\,d\tau \simeq \frac{1}{3}\,\Delta\tau(y_0 + 4y_1 + 2y_2 + \ldots + 4y_{n-1} + y_n)$$

The expression $y(\tau)$ is given by $p(\tau)\cos\omega\tau$ or $p(\tau)\sin\omega\tau$ from Equation (3.82). During a dynamic analysis the response history of a system is required rather than the response at a particular time. Equation (3.82) may, therefore, be written in incremental form using Simpson's rule for each interval $2\Delta\tau$, thus

$$x(t) = \left\{\frac{\Delta\tau}{3m\omega}\left[\sum_0^{t-2\Delta\tau}\!\!\!{}_{\cos} y(\tau) + p(t - 2\Delta\tau)\cos\omega(t - 2\Delta\tau)\right.\right.$$

$$\left.\left. + 4p(t - \Delta\tau)\cos\omega(t - \Delta\tau) + p(t)\cos\omega t\right]\right\}\sin\omega t$$

$$- \left\{\frac{\Delta\tau}{3m\omega}\left[\sum_0^{t-2\Delta\tau}\!\!\!{}_{\sin} y(\tau) + p(t - 2\Delta\tau)\sin\omega(t - 2\Delta\tau)\right.\right.$$

$$\left.\left. + 4p(t - \Delta\tau)\sin\omega(t - \Delta\tau) + p(t)\sin\omega t\right]\right\}\cos\omega t \qquad (3.83)$$

The arithmetical summations

$$\sum_0^{t-2\Delta\tau}\!\!\!{}_{\cos} y(\tau) \quad \text{and} \quad \sum_0^{t-2\Delta\tau}\!\!\!{}_{\sin} y(\tau) \quad \text{are determined}$$

at the preceding increment $(\tau - 2\Delta\tau)$. The terms in square brackets remain constant when the dynamic load has terminated and become the constants in Equation (3.82) for the subsequent free vibration response. The procedure to obtain the response of a SDOF system using numerical integration is fully explained in the example which follows. The tabular form of the calculation may be conveniently carried out using a hand calculator.

Example 3.6

The tower shown in Fig. E3.4 supports a tank and may be idealized as a SDOF system. It is subjected to a half sine pulse loading for a duration of 0.075 second. The response history of the tower is required during the application of the pulse and for the subsequent free vibrations.

The frequency and period of vibration are

$$\omega = \sqrt{\left(\frac{k}{m}\right)} = \sqrt{\left(\frac{1.6 \times 10^6}{4\,000}\right)} = 20\,\text{rad/s}$$

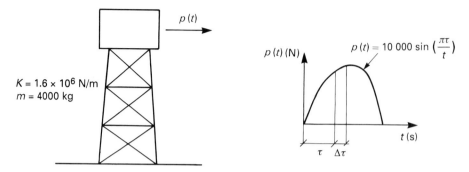

Fig. E3.4   A tower subjected to a half-sine pulse.

and

$$T = \frac{2\pi}{\omega} = 0.314\,\text{s}$$

An arbitrary rule which gives satisfactory results selects the time interval according to $\Delta\tau = T/10$. For the current example, therefore, $\Delta\tau = 0.0075\,\text{s}$.

Table 3.1 illustrates a suitable arrangement to determine the response for the duration of the pulse. For the purpose of the table, Equation (3.83) may be written as

$$x(t) = \frac{\Delta\tau}{3m\omega} (c_1 \sin \omega t - c_2 \cos \omega t) \tag{3.84}$$

and Table 3.1 becomes more or less self-explanatory. Column 6 is the ordinate multiplier and columns 8 and 9 represent the incremental and updated values of c_1 respectively. The values in column 8 are obtained by adding the three rows in column 7 which correspond to an interval $2\Delta\tau$. The values in column 9 are the accumulated sum of those in column 8 and represent updated values of c_1. Columns 10 to 14 may be similarly explained. Column 18 is obtained by multiplying column 17 by $\Delta\tau/3m\omega = 3.125 \times 10^{-8}$.

The elastic restoring forces $f(\tau)$ in the last column are obtained from the product of the displacement $x(\tau)$ and the stiffness k.

Once the loading has terminated, the subsequent free vibrations may be obtained from Equation (3.84) in which c_1 and c_2 have become constants, i.e.

$$x(t) = 3.125 \times 10^{-8} (132\,428 \sin \omega t - 123\,174 \cos \omega t)$$

and the maximum amplitude of motion is given by

$$x_{\text{max}} = \frac{\Delta\tau}{3m\omega} (c_1^2 + c_2^2)^{1/2} = 0.0057\,\text{m}$$

Table 3.1 Undamped response of a tower using numerical integration of the Duhamel integral.

τ (s) 1	P(τ) (N) 2	cos ωτ 3	sin ωτ 4	5 =2×3	6	7 =5+6	8	9	10 =2×4	11	12 =10+11	13	14	15 =9×4	16 =14×3	17 =15×16	x(τ) (m×10⁻⁸) 18	f(τ) (N) 19
				c_1					c_2									
0	0	1.000	0	0	1	0	—	0	0	1	0	—	—	0	0	0	0	0
0.0075	3090	0.989	0.149	3056	4	12224	17837	—	460	4	1840	3580	—	—	—	—	—	—
0.0150	5878	0.955	0.296	5613	1	5613	—	17837	1740	1	1740	—	3580	5280	3419	1861	5816	93
0.0225	8090	0.900	0.432	7281	4	29124	42584	—	3495	4	13980	21094	—	—	—	—	—	—
0.0300	9511	0.825	0.565	7847	1	7847	—	60421	5374	1	5374	—	24674	34138	20356	13782	43069	689
0.0375	10000	0.732	0.680	7320	4	29280	43043	—	6800	4	27200	40021	—	—	—	—	—	—
0.0450	9511	0.622	0.783	5916	1	5916	—	103464	7447	1	7447	—	64605	81012	40240	40772	127413	2039
0.0525	8090	0.497	0.866	4021	4	16084	24128	—	7006	4	28024	40949	—	—	—	—	—	—
0.0600	5878	0.362	0.932	2128	1	2128	—	127592	5478	1	5478	—	105644	118916	38243	80673	252103	4034
0.0675	3090	0.219	0.975	677	4	2708	4836	—	3013	4	12052	17530	—	—	—	—	—	—
0.0750	0	0.071	0.997	0	1	0	—	132428	0	1	0	—	123174	132031	8745	123286	385269	6164

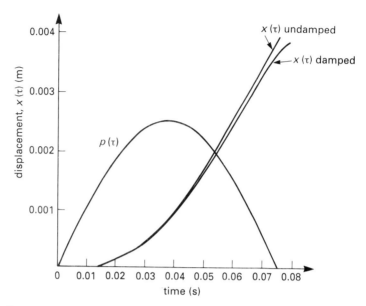

Fig. E3.5 Damped and undamped response of tower.

Figure E3.5 shows the undamped response of the tower during the loading. The free vibration response may be obtained by plotting Equation (3.83).

3.7.2 Damped response

The derivation of the Duhamel integral for a damped system subjected to an arbitrary loading is identical to the undamped analysis except that the free vibrations initiated by the impulse $p(\tau)\,d\tau$ are damped by an exponential decay factor. For this case, the damped equivalent of Equation (3.81) becomes

$$x(t_1) = \frac{1}{m\omega_D} \int_0^t p(\tau)\, e^{-\xi\omega(t-\tau)} \sin \omega_D(t - \tau)\, d\tau \qquad (3.85)$$

where ξ is the damping ratio and ω_D is the damped frequency. Equation (3.85) may be evaluated numerically and once more the Simpson's scheme can be adopted. Equation (3.83) becomes, for the case of a damped system,

$$x(t) = \left\{ \frac{\Delta\tau}{3m\omega} \left[\left(\sum_0^{t-2\Delta\tau} \frac{y(\tau)}{\cos} + p(t - 2\Delta\tau)\cos \omega_D(t - 2\Delta\tau) \right) e^{-\xi\omega 2\Delta\tau} \right. \right.$$

$$\left. + 4p(t - \Delta\tau)\cos \omega_D(t - \Delta\tau)e^{-\xi\omega 2\Delta\tau} + p(t)\cos \omega_D t \right] \right\} \sin \omega t$$

$$- \left\{ \frac{\Delta\tau}{3m\omega} \left[\left(\sum_0^{t-2\Delta\tau} \frac{y(\tau)}{\sin} + p(t - 2\Delta\tau)\sin \omega_D(t - 2\Delta\tau) \right) e^{-\xi\omega 2\Delta\tau} \right. \right.$$

$$+ \; 4p(t - \Delta\tau)\sin\omega_D(t - \Delta\tau)e^{-\xi\omega2\Delta\tau} + p(t)\sin\omega_D t\Big]\Big\}\cos\omega t$$

$$(3.86)$$

Example 3.7

The damped response of the tower which has been considered in the previous example has been obtained in Table 3.2 for a damping ratio $\xi = 0.067$. The arrangement for the table is similar to the previous example, but with the following modifications. The terms $e^{-\xi\omega\Delta\tau}$ and $e^{-\xi\omega2\Delta\tau}$ are combined with the ordinate multiplying factors in columns 6 and 11, i.e.

$$e^{-\xi\omega2\Delta\tau} = e^{-0.02} = 0.98$$

$$4e^{-\xi\omega\Delta\tau} = 4e^{-0.01} = 3.96$$

Since the system is lightly damped, the damped natural frequency is assumed to be identical to the undamped value. The decay term $e^{-\xi\omega2\Delta\tau}$ acts on the updated value of c_1 as well as the appropriate part of the new increment and these two parts are added together before multiplying by the decay term. For example, consider the first interval, $2\Delta\tau = 0.015$, in Table 3.2

$$\text{column } 9 = 0 + 12\,102 + 5\,613 = 17\,715$$
$$\text{column } 7 = 17\,715 + 5\,613 = 23\,328$$
$$\text{column } 8 = 23\,328 \times 0.98 = 22\,861$$

Columns 14, 12 and 13 are obtained in a similar manner. The calculation of $x(\tau)$ and $f(\tau)$ follows as before for the undamped response. The damped response of the tower during the loading is shown in Fig. E3.5.

3.8 Systems with rigid body components

It has been assumed for the previously considered SDOF systems that the mass, damping and stiffness have been concentrated at discrete points. This idealization may not be sufficient for those SDOF systems in which the mass is significantly distributed throughout the system. In such cases distributed inertia forces occur as a result of rigid component accelerations. Finite beams and plates which are constrained to move as SDOF systems by springs and dampers are examples for which the rigid-body inertia effects must be considered. For simplicity, elastic and damping forces in rigid-body systems are assumed to be concentrated at the centre of mass of the rigid-body component.

Figure 3.12 provides some values of mass and moment of inertia for common rigid-body components. The inclusion of distributed inertia forces in the equation of motion should not cause great difficulty and the examples which follow illustrate the manner in which this is achieved.

Table 3.2 Damped response of a tower using numerical integration of the Duhamel integral.

				c_1					c_2									
τ (s)	$P(\tau)$ (N)	$\cos\omega\tau$	$\sin\omega\tau$	= 2×3		= 5+9			= 2×4		= 10+14			= 9×4	= 14×9	= 15×16	$x(\tau)$ (m×10⁻⁸)	$f(\tau)$ (N)
1	2	3	4	5	6	7	8	9	10	11	12	13	14	15	16	17	18	19
0	0	1.000	0	0	0.98	0	0	0	0	0.98	0	0	0	0	0	0	0	0
0.0075	3030	0.989	0.149	3056	3.56	—	12102	—	460	3.96	—	1822	—	—	—	—	—	—
0.0150	5878	0.955	0.296	5613	0.98	23328	22861	17715	1740	0.98	5302	5196	3562	5244	3402	1842	5756	92
0.0225	8090	0.900	0.432	7281	3.96	—	28833	—	3495	3.96	—	13840	—	—	—	—	—	—
0.0300	9511	0.825	0.565	7847	0.98	67388	66040	59541	5374	0.98	29784	29188	24410	33641	20138	13503	42196	675
0.0375	10000	0.732	0.680	7320	3.96	—	28987	—	6800	3.96	—	26928	—	—	—	—	—	—
0.0450	9511	0.622	0.783	5916	0.98	106859	104722	100943	7447	0.98	71010	69588	63563	79038	39536	39502	123444	1975
0.0525	8090	0.497	0.866	4021	3.96	—	15923	—	7006	3.06	27744	—	—	—	—	—	—	—
0.0600	5878	0.362	0.932	2128	0.98	124901	122403	122773	5478	0.98	108288	106122	102810	114424	37217	77153	241103	3858
0.0675	3090	0.219	0.975	677	3.96	—	2681	—	3013	3.96	—	11931	—	—	—	—	—	—
0.0750	0	0.071	0.997	0	0.98	—	0	125054	0	0.58	—	—	118053	124709	8382	116347	363584	5817

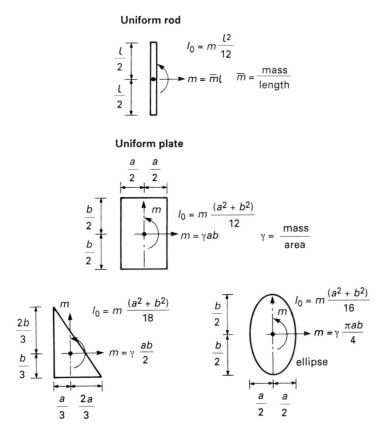

Fig. 3.12   Rigid-body mass and mass moment of inertia.

Example 3.8

A cantilever beam is hinged at A and is supported by a spring of stiffness k, at B (Fig. E3.6). If the mass of the cantilever is m per unit length, calculate the natural frequency of the system when the spring is displaced a small amount and suddenly released.

The beam is constrained to oscillate about the hinge at A as a rigid body and will perform undamped motion. If the spring displacement is assumed small and equal to $x(t)$, the associated spring force is $kx(t)$ and the moment of this force about A is $kx(t)l$.

The usual arrangements of SDOF rigid systems, which are familiar to the structural engineer, may be analyzed by taking moments about some point other than the centre of mass of the rigid components of the system. Here it is assumed that the reader has some knowledge of the kinematics of a rigid body in plane motion and it will be stated without proof that the moment of a rigid body about some point other than its centre of mass is obtained by summing the moment of inertia about the centre of mass times the angular

Fig. E3.6   Rigid beam example.

acceleration and a term which is equal to the moment of the total mass times the acceleration of the centre of the mass. In this way the forces due to the distributed mass are accounted for.

For the current problem, the equation of motion is obtained by taking moments about A for all the forces in the system. The moment of inertia of the rigid beam is $ml^2/12$ and for small displacements the angular acceleration is $\ddot{x}(t)/l$. The acceleration of the centre of mass of the beam is $\ddot{x}(t)/2$. Thus,

$$\frac{ml^2}{12}\frac{\ddot{x}(t)}{l} + ml\,\frac{\ddot{x}(t)}{2}\frac{l}{2} + kx(t)l = 0$$

and the equation of motion becomes

$$\frac{ml}{3}\ddot{x}(t) + kx(t) = 0$$

from which

$$\omega = \sqrt{\frac{3k}{ml}} \text{ rads/s} \qquad \text{or} \qquad f = \frac{1}{2\pi}\sqrt{\frac{3k}{ml}} \text{ hertz}$$

3.9 Structural damping

For the SDOF systems which have so far been considered, the energy loss mechanism has been represented by a viscous damper for which the damping force resisting motion has been assumed to be proportional to the velocity of the system. It has been previously stated that this is a particularly convenient representation enabling the equations of motion to be solved without difficulty.

The damping for engineering structures is usually more complex than a viscous idealization (although an equivalent viscous quantity may be evaluated) and may be caused by different sources. Additionally, it is relatively straightforward to evaluate the mass and stiffness of a structure, but much more difficult to assess the amount of damping which is present. In many instances it is usual to refer to experimental procedures in order to estimate structural damping.

A further complication which the engineer is likely to encounter concerns the actual definition of damping. There is no generally accepted definition or set of symbols describing damping and when damping information is being interpreted great care should be taken to ensure that the correct factors are being considered.

3.9.1 Damping definitions

Some of the more common definitions of damping will now be considered and, for completeness, these will include the definitions which have already been encountered in this chapter.

Critical damping coefficient, c_c, is the least value of the damping coefficient, c, required to prevent oscillation of the system and is case (b) of section 3.2.2, where $c_c = 2m\omega$. Although c and c_c are termed coefficients, they have dimensions of force/velocity.

Damping ratio (or factor), ξ, is dimensionless and relates the actual damping coefficient to the critical damping coefficient and is defined by Equation (3.17).

Logarithmic decrement, δ, is the natural logarithm of the ratio of successive amplitudes of free vibration one complete cycle apart, i.e. Equation (3.26):

$$\delta = \ln \left(\frac{x_n}{x_{n+1}}\right) = 2\pi\xi\left(\frac{\omega}{\omega_D}\right)$$

Equation (3.26) relates the logarithmic decrement to the damping ratio. For small values of damping, $\delta = 2\pi\xi$, which is obtained from Equation (3.27). This approximation is only 2% in error for a value of ξ of 0.20 ($\delta \approx 1.25$) and is considered to be a relatively high degree of damping.

For non-successive amplitudes of vibration m cycles apart, the logarithmic decrement is given by Equation 3.30:

$$\ln \left(\frac{x_n}{x_{n+m}}\right) = 2\pi m\xi\left(\frac{\omega}{\omega_D}\right)$$

The term 'decay factor' is used to denote the logarithmic decrement divided by period of vibration.

Solid damping factor, γ, is related to the logarithmic decrement as follows

$$\gamma = \frac{\delta}{\pi} = 2\zeta \tag{3.87}$$

for small values of the damping ratio.

Damping capacity. This term is used to specify the damping in metals at various stress levels and is usually expressed as the percentage loss of energy per cycle obtained from a torsional vibration test. A 1% loss per cycle is equivalent to a logarithmic decrement of about 0.005; 5% corresponds to $\delta = 0.025$ and 30% to $\delta = 0.18$.

Total damping energy, D_0, is the energy dissipated in a body during one cycle of vibration and its dimensions are Nm/cycle. The total damping energy can be calculated from the specific damping energy, D, by an integration over the entire body

$$D_0 = \int_v D \, dv \tag{3.88}$$

Specific damping energy, D, is the energy dissipated at a particular point in the volume of a specimen during one cycle of vibration and has the same units as the total damping energy. It is the most fundamental of all the known definitions of damping since it is dependent only on the material subject to investigation and is not influenced by the shape or volume of the specimen or the stress distribution. The damping ratio for an entire specimen may be related to the total damping energy as follows

$$\xi = \frac{D_0}{4\pi W_0} \tag{3.89}$$

where W_0 is the strain energy of the specimen.

3.9.2 Damping models

In practice, all engineering systems possess damping and are, therefore, called non-conservative systems since energy is being dissipated by damping forces. The effect of damping is to increase the period of natural frequency and to make the resonant frequency somewhat less than the value obtained without damping being present.

It has been stated previously that it is relatively straightforward to define the mass and stiffness characteristics of a system, however, the damping characteristics are less understood. Damping may be deliberately introduced using energy transfer devices but these will not be considered since they are beyond the scope of this book. For convenience, damping is classified as system damping or material damping and alternative models have been proposed to describe the nature of damping within this general classification.

Coulomb damping. System damping involves energy loss mechanisms between distinguishable parts of a structure. The most relevant system damping mechanism of concern to the structural engineer is Coulomb damping, which results from the motion of a body on a dry surface. Since bolted connections and similar jointing arrangements are a common constructional feature, energy can be dissipated during cyclic shear strain of the dry surfaces of the connection. The resultant damping is nearly consistent between the moving parts and depends upon the normal pressure, N, and the coefficient of kinetic friction, μ, i.e. 'damping force' equals μN. The successive amplitudes of the free vibration for a system with Coulomb damping decay linearly.

Viscous damping, which has already been considered, is the commonest form of system damping used in dynamic analysis. Lubricated surfaces, shock absorbers and dashpots are examples where viscous damping may be used to represent these energy dissipating effects. The use of this type of damping in the analysis of the dynamic behaviour of structural systems is particularly convenient since it enables the governing equations of motion to be solved conveniently. Viscous damping forces are proportional to velocity and successive amplitudes of vibration decay exponentially.

Hysteretic damping is a form of material damping and is a fundamental material property. It is particularly appropriate for structural materials for which experimental evidence suggests that damping is frequency independent rather than frequency dependent, which is the case for the viscous damping concept. The damping force associated with hysteretic damping is in phase with the velocity, but proportional to the displacement of a system. The hysteretic damping force may be expressed as

$$f_D = \zeta k \left(\frac{f_s}{k}\right)\left(\frac{\dot{x}}{|\dot{x}|}\right) \tag{3.90}$$

where ζ is the hysteretic damping coefficient. The hysteretic damping force is expressed as a fraction of the elastic force.

3.9.3 Experimental determination of damping

It has been stated previously that it is often necessary to determine the value of damping in an elastic system either by free vibration tests or by subjecting the system to a forced vibration and observing the response for a range of frequencies at and in the region of resonance.

In the former case, if successive amplitudes of free vibration are known, then the logarithmic decrement of damping, δ, may be obtained from Equation (3.26) or (3.30) in which ω equals ω_D. Since structural damping is usually small, it is more accurate to use Equation (3.30) for amplitudes which are several cycles apart. The equipment and instrumentation to be used with the free vibration method are minimal and the input causing the free vibrations may be induced by any convenient means. For instance, the damping characteristics of tall buildings have been obtained by observing the free vibration response caused by wind, non-damaging impact or even the release of a taut cable connecting the building to the ground. Only relative displacements need to be measured during free vibration.

In order to determine the damping ratio, ξ, from forced vibration tests, it is necessary to excite the system using a steady-state harmonic source. Suitable devices for generating steady-state vibrations are discussed in section 6.8.1 and a typical frequency response curve for a SDOF system excited by such a device is shown in Fig. 3.13. From this curve, it is possible to obtain the dynamic magnification factor for any given frequency. This factor which is the ratio of the response amplitude at a given frequency to the static displacement (response at zero frequency), is given by Equation (3.45). At resonance, when β and $\cos(\omega t - \alpha)$ equal unity, it can be seen that Equation (3.45) may be rearranged as follows

$$\xi = \frac{1}{2} \frac{x_s}{x(t)} \qquad (3.91)$$

Measurements which are required at or near to the resonant response of a system are often difficult to obtain and the half power or bandwidth method of determining the damping ratio is to be preferred. In this method, the frequencies of response for the power input at resonance are determined. These frequencies correspond to a response amplitude of $1/\sqrt{2}$ times the

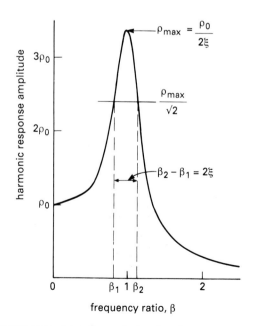

Fig. 3.13   Frequency-response curve for a damped system.

resonant amplitude. By equating the peak amplitude of the steady-state response given by Equation (3.43) to $1/\sqrt{2}$ times $x(t)$ from Equation (3.45), two half power frequency ratios β_1 and β_2 are obtained. It can be shown that the damping ratio is given by half the difference in these values, i.e.

$$\xi = \frac{1}{2}(\beta_2 - \beta_1) \tag{3.92}$$

Figure 3.13 shows that a horizontal line may be drawn through the response curve at $1/\sqrt{2}$ times the resonant amplitude. The intersections of this line with the curve provide the values of β_1 and β_2 for Equation (3.92).

3.9.3.1 *Equivalent viscous damping*

Non-viscous damping in a system may be reduced to an equivalent amount of viscous damping by measuring the forced response at resonance. The response of a SDOF system at resonance is 90° out of phase with the input and the damping force is balanced by the applied loading. By using equipment which is capable of detecting phase difference, the resonant condition may be established by adjusting the frequency of the input. If the applied load is plotted against the displacement for one cycle then the damping force−displacement relationship is automatically obtained.

For systems with viscous damping, the damping force−displacement relationship is represented as an ellipse (Fig. 3.14) and the damping ratio, c, is defined as the ratio of the maximum damping force, f_D, to the maximum velocity, ωx_{max}, i.e.

$$c = \frac{f_D}{\omega x_{max}} \tag{3.93}$$

For systems with non-viscous damping, the relationship is no longer

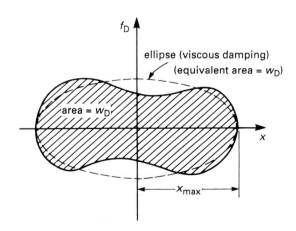

Fig. 3.14   Viscous and non-viscous damping.

represented by an ellipse and a curve such as that shown in Fig. 3.14 may be obtained. An equivalent amount of viscous damping may be defined by an ellipse with the same maximum displacement and area as the curve representing non-viscous damping. If it is assumed that the energy lost per cycle is the same for viscous and non-viscous damping then an equivalent viscous damping coefficient is given by

$$C_{EVD} = \frac{A_{EVD}}{\omega \pi x_{max}^2} \tag{3.94}$$

where A_{EVD} is the area of the equivalent viscous damping curve and equals $\pi f_D x_{max}$ for the ellipse.

The critical damping coefficient, c_c, is given by $2m\omega$ or $2k/\omega$, for a SDOF system. The stiffness k may be obtained from the force–displacement relationship for the system. If A is the area beneath the force–displacement curve then the stiffness $k = 2A/x_{max}^2$ for a linear system. The damping ratio for an equivalent viscous system is, therefore, given by

$$\xi = \frac{c}{c_c} = \frac{A_{EVD}}{4\pi A} \tag{3.95}$$

Since Equation (3.94) includes a frequency term, the damping ratio of Equation (3.95) is also frequency dependent.

Hysteretic damping effects may be represented by an equivalent viscous damper. A typical hysteretic damping force–displacement relationship is shown in Fig. 3.15. Since the hysteretic damping force is expressed as a fraction of the elastic force, the energy lost per cycle is $2\zeta kx_{max}^2$ or $4\zeta A$ and from Equation (3.95)

$$\zeta = \pi \xi \tag{3.96}$$

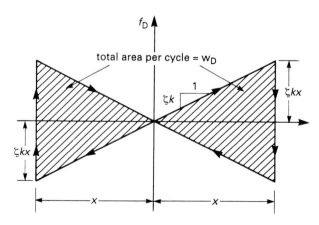

Fig. 3.15 Hysteretic damping.

3.9.4 Damping in structural materials

Damping in most metals is very small and is much less than in other materials such as cork, rubber and concrete. *Kempe's Engineer's Yearbook* (published annually) provides some information on the damping properties of metals. Typical values of metal damping capacity ranging from 5% to 30% are given (corresponding logarithmic decrements are 0.025 to 0.18).

The improved damping properties of rubber and cork are often used to control unnecessary vibration. Manufacturers usually provide data relating to the damping characteristics of their products. Values of logarithmic decrements ranging from 0.01 to 0.50 are quoted; particular values depending upon the constituents of the rubber. Cork would appear to resemble a medium hard rubber for which logarithmic decrements in the range 0.2 to 0.5 have been obtained.

The damping properties of concrete depend upon whether the material is cracked or uncracked. Mean values for the logarithmic decrement of concrete have been quoted in the order of 0.03 (uncracked) and 0.1 (cracked). The constituents of the concrete and the amount and type of reinforcement influence the amount of damping present. It is accepted that the amount of damping in prestressed concrete is less than in reinforced or plain concrete. Values of logarithmic decrement have been quoted for prestressed and reinforced concrete beams in the range of 0.03 to 0.44.

3.9.5 Damping in structures

Attempts have been made to evaluate the damping ratios of complete structures. These ratios represent the overall effect of damping and, therefore, provide an indication of likely levels. The actual value of the damping ratio will vary for different structures, however, the following information should be of interest to enable an assessment of dynamic response to be made.

For concrete buildings, damping ratios in the range 0.05 to 0.10 have been obtained and for masonry buildings, values of 0.05 to 0.15 have been quoted. In particular cases, the damping ratio for a three-storey building with a reinforced concrete frame has been found to be 0.02 to 0.03 and for an eleven-storey reinforced concrete framed structure, previously damaged by an earthquake, the damping ratio was found to be 0.05 to 0.06.

The damping in other types of structure has been obtained, for instance, a value for the logarithmic decrement of 0.193 for a concrete cooling tower has been quoted. Finally, the damping capacity for a metal bridge has been found to be 4% to 6% for frequencies of 1 to 3 hertz and 10% to 20% for frequencies of 4 to 10 hertz.

Chapter Four
Systems With Multiple Degrees of Freedom

4.1 Introduction

The essential properties of a structure which is to be idealized as a SDOF system are that its mass, stiffness, damping and loading may be concentrated in such a way that the motion can be described by a single co-ordinate or mode of vibration. For structures in which these properties are distributed, the SDOF idealization is inadequate and the results derived from the associated analysis will be unreliable. The motion of such structures must be described by more than one co-ordinate or mode of vibration.

Although a large number of degrees of freedom are usually associated with a complex structural system, acceptable results may be obtained from the analysis of the response of only a few degrees of freedom. Some simple examples of a MDOF beam system have been given in section 1.3 of Chapter 1.

This chapter is concerned with the analysis of MDOF systems and, in particular, a matrix approach is adopted. A formal analysis is presented initially, when it will be shown that the dynamic analysis of a structural system is a form of the classical eigenvalue problem of matrix algebra. The methods of analysis which have been proposed towards the end of the chapter are suitable as a basis for design office calculations.

4.2 Equations of motion for a system with two degrees of freedom

The simple two-mass beam which has been considered in Chapter 1 will be used to obtain the equations of motion for a 2DOF system. Figure 4.1 shows the arrangement of the beam and the forces acting on each mass. It is assumed that each mass is constrained to move only in the vertical plane. The associated displacements therefore represent the two independent co-ordinates or degrees of freedom which will be used to define the configuration of the system.

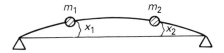

Fig. 4.1   Two degrees of freedom system.

The equations of motion can be obtained by considering the dynamic equilibrium of each mass in turn. For a general situation, four types of force usually act upon each mass and these are the applied, inertia, damping and elastic forces respectively. The elastic force acting on a mass depends not only upon the displacement of the mass under consideration, but also upon the displacement of the adjacent mass. This is the only coupling which is allowed in the lumped parameter idealizations to be considered in this chapter. The remaining three forces acting on a mass are, therefore, assumed to be independent of the motion of the adjacent mass. The equations of dynamic equilibrium for each mass may now be written as:

$$m_1\ddot{x}_1 + c_1\dot{x}_1 + k_{11}x_1 + k_{12}x_2 = p_1(t) \tag{4.1}$$

for mass m_1, and

$$m_2\ddot{x}_2 + c_2\dot{x}_2 + k_{21}x_1 + k_{22}x_2 = p_2(t) \tag{4.2}$$

for mass m_2.

Equations (4.1) and (4.2) may be combined into a single matrix equation as follows

$$\begin{bmatrix} m_1 & 0 \\ 0 & m_2 \end{bmatrix} \begin{Bmatrix} \ddot{x}_1 \\ \ddot{x}_2 \end{Bmatrix} + \begin{bmatrix} c_1 & 0 \\ 0 & c_2 \end{bmatrix} \begin{Bmatrix} \dot{x}_1 \\ \dot{x}_2 \end{Bmatrix} + \begin{bmatrix} k_{11} & k_{12} \\ k_{21} & k_{22} \end{bmatrix} \begin{Bmatrix} x_1 \\ x_2 \end{Bmatrix} = \begin{Bmatrix} p_1(t) \\ p_2(t) \end{Bmatrix} \tag{4.3}$$

or

$$\mathbf{M\ddot{X} + C\dot{X} + KX = P(t)} \tag{4.4}$$

where the matrix containing the masses m_1 and m_2 is called the mass matrix, and the matrix containing the damping coefficients c_1 and c_2 is called the damping matrix. The matrix containing the coefficients k_{11}, k_{12}, etc. is the familiar linear elastic stiffness matrix for the beam. Each coefficient in the stiffness matrix is a force which is caused by a unit displacement, i.e.

(k_{ij}) = the force at position i due to a unit displacement at position j when all other displacements are zero

The vectors containing the accelerations, velocities, displacements and applied time-dependent loading are given by $\mathbf{\ddot{X}}$, $\mathbf{\dot{X}}$, $\mathbf{X}$ and $\mathbf{P(t)}$ respectively.

4.3 Equations of motion for a general system with multiple degrees of freedom

If the beam of Fig. 4.1 had n masses instead of the two which are shown, an equation of dynamic equilibrium for the ith mass in the system can be obtained. This equation is similar to Equations (4.1) and (4.2), thus

$$m_i\ddot{x}_i + c_i\dot{x}_i + k_{i1}x_1 + k_{i2}x_2 + \ldots + k_{ii}x_i + \ldots + k_{in}x_n = p_i(t) \quad (4.5)$$

The inertia, damping and applied force terms relate only to the ith mass. The elastic force acting on the ith mass has contributions from not only the x_ith displacement but from the displacements of all other masses. Since an equation similar to Equation (4.5) may be written for each mass in turn, a set of n linear simultaneous equations can be obtained for the system and these may be written in the matrix notation of Equation (4.4). The matrices and column vectors within Equation (4.4) become

$$\mathbf{M\ddot{X}} = \begin{bmatrix} m_1 & 0 & . & . & . & 0 \\ 0 & m_2 & & & & \\ . & & . & & & \\ . & & & m_i & & \\ . & & & & . & \\ 0 & & . & . & & m_n \end{bmatrix} \begin{Bmatrix} \ddot{x}_1 \\ \ddot{x}_2 \\ . \\ \ddot{x}_i \\ . \\ \ddot{x}_n \end{Bmatrix} \quad (4.6)$$

$$\mathbf{C\dot{X}} = \begin{bmatrix} c_1 & 0 & . & . & . & 0 \\ 0 & c_2 & & & & \\ . & & . & & & \\ . & & & c_i & & \\ . & & & & . & \\ 0 & & . & . & . & c_n \end{bmatrix} \begin{Bmatrix} \dot{x}_1 \\ \dot{x}_2 \\ . \\ \dot{x}_i \\ . \\ \dot{x}_n \end{Bmatrix} \quad (4.7)$$

$$\mathbf{KX} = \begin{bmatrix} k_{11} & k_{12} & . & . & . & k_{1n} \\ k_{21} & k_{22} & & & & \\ . & & . & & & \\ . & & & k_{ii} & & \\ . & & & & . & \\ k_{n1} & & . & . & . & k_{nn} \end{bmatrix} \begin{Bmatrix} x_1 \\ x_2 \\ . \\ x_i \\ . \\ x_n \end{Bmatrix} \quad (4.8)$$

$$\mathbf{P(t)} = \begin{Bmatrix} p_1(t) \\ p_2(t) \\ . \\ p_i(t) \\ . \\ p_n(t) \end{Bmatrix} \quad (4.9)$$

Equation (4.4) is the MDOF equivalent of Equation (3.1) for a SDOF and Equations (4.6) to (4.9) represent the inertia, damping, elastic and applied forces acting on the masses in an n degree of freedom beam system. This system is a particular type of lumped parameter idealization, which could be used to obtain the dynamic response of a beam; the distributed mass being represented by a series of concentrated points.

In section 1.4 of Chapter 1 some general comments were made regarding lumped mass idealization. The 4DOF building frame shown in Fig. 1.3 could have been used to obtain four equations of motion similar to Equations (4.1) to (4.4) and hence Equations (4.5) to (4.9). Some reinterpretation of the notation would be necessary, but the equations would be identical in form. The mass matrix would represent the concentrated masses at each storey. The corresponding damping and applied force for each storey would be represented by the damping matrix and applied load vector respectively. Since the frame is constrained to move in the horizontal direction the displacement, velocity and acceleration vectors correspond accordingly for each storey. The stiffness matrix will be obtained from a rigidly jointed plane frame constrained to move horizontally, i.e. only flexural deformation of the vertical members in the plane of the frame is permitted.

In the same way that the analysis for the two-mass beam system was extended to include n masses, so the analysis for the 4DOF shear frame building may be extended to include n storeys. Clearly, whenever it is considered pertinent to idealize a structure as a lumped parameter system, a set of simultaneous equations will be obtained which can be arranged in the form of Equation (4.4). The number of equations so formed equals the total number of idealized degrees of freedom for the structure.

4.4   Free vibration

4.4.1   Frequency analysis

Equation (4.4) can be modified to represent the equations of motion for an undamped system undergoing free vibrations in any one of its possible modes of vibration. For this case the damping and applied forces are omitted to give

$$\mathbf{M\ddot{X} + KX = 0} \tag{4.10}$$

where the right-hand side is a column vector containing only zeroes. It should be noted that Equation (4.10) is analogous to Equation (3.2) when $c\dot{x}(t)$ equals zero.

It has been shown in Chapter 3 that the free vibration response of a SDOF system is simple harmonic motion for which $\ddot{x} = -\omega^2 x$. It may be assumed that a similar relationship is valid for the free vibration response of MDOF systems, therefore

$$\ddot{\mathbf{X}} = -\omega^2 \mathbf{X} \tag{4.11}$$

Substituting Equation (4.11) into Equation (4.10) gives

$$-\omega^2 \mathbf{MX} + \mathbf{KX} = 0$$

or   (4.12)

$$|\mathbf{K} - \omega^2 \mathbf{M}|\ \mathbf{X} = 0$$

Equations (4.12) form a set of simultaneous equations in $\mathbf{X}$ with ω^2 as the unknowns. The particular way in which these equations of motion have been formulated in matrix notation represents what is generally called an eigenvalue problem in matrix algebra.

Cramer's rule which is used to solve Equation (4.12) requires, for a non-trivial solution, that the determinant of the coefficients of $\mathbf{X}$ equals zero, i.e.

$$|\mathbf{K} - \omega^2 \mathbf{M}| = 0 \tag{4.13}$$

Equation (4.13) is called the frequency equation of the system. By expanding this determinant into its co-factors an equation of the nth degree in ω^2 is obtained for a system having n degrees of freedom. The values of ω^2 which satisfy this equation, i.e. $\omega_1, \omega_2, \ldots \omega_n$, are called eigenvalues and correspond to the frequencies of the n nodes of vibration at which the system can vibrate. It is usual to let the smallest frequency be represented by ω_1. This frequency is termed the fundamental frequency of the system and corresponds to the first mode of vibration. The remaining frequencies correspond to the second and higher modes of vibration. The column vector $\mathbf{X}$ which corresponds to a particular mode of vibration is called an eigenvector.

Equation (4.12) and hence Equation (4.13) require the stiffness matrix, $\mathbf{K}$, of the structure to be obtained. In many cases it is more convenient to obtain the flexibility matrix, $\mathbf{F}$, of a structure. Since the stiffness matrix is the inverse of the flexibility matrix, Equation (4.12) is easily modified by multiplying throughout by $\mathbf{F}/\omega^2$, to give

$$\left[\frac{1}{\omega^2}\mathbf{I} - \mathbf{FM}\right]\mathbf{X} = 0 \tag{4.14}$$

where $\mathbf{I}$ is an identity matrix. The frequency equation is now

$$\left[\frac{1}{\omega^2}\mathbf{I} - \mathbf{FM}\right] = 0 \tag{4.15}$$

The previous considerations for Equation (4.13) are applicable except that the eigenvalues are given by $1/\omega^2$ instead of ω.

4.4.2 Mode shape analysis

Having obtained the frequencies of vibration of the modes, it is necessary to define the shape of each mode by determining the eigenvector **X**. Since Equations (4.12) and (4.14) are satisfied identically and the eigenvalues already obtained, the actual values of the displacement in the eigenvector are indeterminant. It is possible, however, to determine the shape of a mode by assigning an arbitrary value to a particular displacement and solving for all other displacements relative to the arbitrary value.

If ϕ_i represents the *i*th eigenvector for an *n* degree of freedom system, then the shape of the mode is given by

$$\boldsymbol{\phi}_i = \begin{Bmatrix} \phi_{1i} \\ \phi_{2i} \\ \vdots \\ \phi_{ii} \\ \vdots \\ \phi_{ni} \end{Bmatrix} \tag{4.16}$$

Since there will be *n* modes, the shape of which are given by ϕ_1, ϕ_2, ..., ϕ_n, a square matrix may be written to include the *n* mode shapes, i.e.

$$\boldsymbol{\phi} = \begin{bmatrix} \phi_{11} & \cdot & \cdot & \phi_{1i} & \cdot & \cdot & \phi_{1n} \\ \phi_{21} & \cdot & \cdot & \phi_{2i} & \cdot & \cdot & \phi_{2n} \\ \cdot & & & \cdot & & & \cdot \\ \phi_{n1} & \cdot & \cdot & \phi_{ni} & \cdot & \cdot & \phi_{nn} \end{bmatrix} \tag{4.17}$$

The procedure for obtaining the frequencies and mode shapes is given in the following example.

Example 4.1

Find the natural frequencies and mode shapes of the three-storey building shown in Fig. E4.1. Assume that the building is undamped.

The building can be idealized as a three-degrees-of-freedom lumped mass system with the mass matrix given by

$$\mathbf{M} = 10^5 \begin{bmatrix} 4 & 0 & 0 \\ 0 & 4 & 0 \\ 0 & 0 & 4 \end{bmatrix} \quad \text{kg}$$

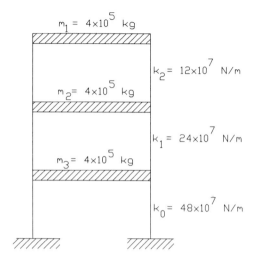

Fig. E4.1

The stiffness matrix can be obtained by assuming unit displacements at each storey and then calculating the forces in the structure. Hence

$$\mathbf{K} = 120 \times 10^6 \begin{bmatrix} 1 & -1 & 0 \\ -1 & 3 & -2 \\ 0 & -2 & 6 \end{bmatrix} \text{ N/m}$$

The natural frequencies can be obtained from the determinant of Equation (4.13)

$$|\mathbf{K} - \omega^2 \mathbf{M}| = 0$$

$$\mathbf{K} - \omega^2 \mathbf{M} = 120 \times 10^6 \begin{bmatrix} (1 - \alpha) & -1 & 0 \\ -1 & (3 - \alpha) & -2 \\ 0 & -2 & (6 - \alpha) \end{bmatrix}$$

where $\alpha = \dfrac{\omega^2}{300}$.

The determinant of the matrix is

$$(1 - \alpha)\,[(3 - \alpha)\,(6 - \alpha) - 2 \times 2] + [-(6 - \alpha) + 0] = 0$$

$$(1 - \alpha)\,(18 - 9\alpha + \alpha^2 - 4) + \alpha - 6 = 0$$

$$\alpha^3 - 10\alpha^2 + 22\alpha - 8 = 0$$

Solutions of this equation are

$$\alpha_1 = 0.45 \qquad \alpha_2 = 2.52 \qquad \alpha_3 = 7.034$$

thus giving

$$\omega^2 = \begin{bmatrix} 135 \\ 756 \\ 210 \end{bmatrix} \quad \text{and} \quad \omega = \begin{bmatrix} \omega_1 \\ \omega_2 \\ \omega_3 \end{bmatrix} = \begin{bmatrix} 11.62 \\ 27.50 \\ 45.94 \end{bmatrix} \quad \text{rad/s}$$

The mode shapes can now be calculated. Assuming a normalized shape for each mode, i.e. $\phi_{1n} = 1$, gives

$$\Phi_n = \begin{bmatrix} 1 \\ \phi_{2n} \\ \phi_{3n} \end{bmatrix}$$

Thus Equation (4.12) becomes

$$120 \times 10^6 \begin{bmatrix} (1 - \alpha) & -1 & 0 \\ -1 & (3 - \alpha) & -2 \\ 0 & -2 & (6 - \alpha) \end{bmatrix} \begin{bmatrix} 1 \\ \phi_{2n} \\ \phi_{3n} \end{bmatrix} = \begin{bmatrix} 0 \\ 0 \\ 0 \end{bmatrix}$$

Partial multiplication of the matrices gives

$$\begin{bmatrix} E_{01}^n \end{bmatrix} + \begin{bmatrix} E_{00}^n \end{bmatrix} \cdot \begin{bmatrix} \phi_{2n} \\ \phi_{3n} \end{bmatrix} = 0$$

where

$$\begin{bmatrix} E_{00}^n \end{bmatrix} = \begin{bmatrix} (3 - \alpha) & -2 \\ -2 & (6 - \alpha) \end{bmatrix} \quad \begin{bmatrix} E_{01}^n \end{bmatrix} = \begin{bmatrix} -1 \\ 0 \end{bmatrix}$$

Thus

$$\begin{bmatrix} \phi_{2n} \\ \phi_{3n} \end{bmatrix} = -\begin{bmatrix} E_{00}^n \end{bmatrix}^{-1} \begin{bmatrix} E_{01}^n \end{bmatrix}$$

Hence the first mode shape is given by

$$\begin{bmatrix} \phi_{21} \\ \phi_{31} \end{bmatrix} = -\begin{bmatrix} (3 - 0.45) & -2 \\ 2 & (6 - 0.45) \end{bmatrix}^{-1} \begin{bmatrix} -1 \\ 0 \end{bmatrix} = \frac{-1}{10.15} \begin{bmatrix} 5.55 & 2 \\ 2 & 2.55 \end{bmatrix} \begin{bmatrix} -1 \\ 0 \end{bmatrix}$$

giving

$$\begin{bmatrix} \phi_{21} \\ \phi_{31} \end{bmatrix} = \begin{bmatrix} 0.547 \\ 0.197 \end{bmatrix} \quad \text{and hence} \quad \Phi_1 = \begin{bmatrix} 1 \\ 0.547 \\ 0.197 \end{bmatrix}$$

Similarly for the second and third modes

$$\begin{bmatrix} \phi_{22} \\ \phi_{32} \end{bmatrix} = \frac{+1}{2.33} \begin{bmatrix} 2.48 & 2 \\ 2 & 0.48 \end{bmatrix} \begin{bmatrix} -1 \\ 0 \end{bmatrix} = \begin{bmatrix} -1.49 \\ -0.86 \end{bmatrix}$$

$$\begin{bmatrix} \phi_{23} \\ \phi_{33} \end{bmatrix} = \frac{-1}{0.171} \begin{bmatrix} -1.034 & 2 \\ 2 & -4.034 \end{bmatrix} \begin{bmatrix} -1 \\ 0 \end{bmatrix} = \begin{bmatrix} -6.05 \\ +11.70 \end{bmatrix}$$

The mode shapes for the three natural frequencies are

$$\phi_1 = \begin{bmatrix} 1 \\ 0.547 \\ 0.197 \end{bmatrix}; \quad \phi_2 = \begin{bmatrix} 1 \\ -1.49 \\ -0.86 \end{bmatrix}; \quad \phi_3 = \begin{bmatrix} 1 \\ -6.05 \\ 11.70 \end{bmatrix}$$

4.5 Orthogonality of mode shapes

It is convenient at this stage to introduce some properties of the free vibration mode shapes which will be extremely useful in subsequent dynamic analysis. These properties are referred to as the orthogonality relationships and will be explained with reference to Fig. 4.2. In this figure a structure has been assumed to vibrate in two possible arbitrary modes. For undamped free vibrations the dynamic equilibrium condition is given by Equation (4.10). This equation may be rearranged so that the inertia forces are equated to the elastic forces. The free vibration motion, therefore, involves elastic deflections caused by inertia effects. The inertia forces may be regarded as static forces deflecting the beam into the same shape as the dynamic mode.

Applying the reciprocal theorem to the structure shown in Fig. 4.2,

$$\omega_i^2 m_1 x_{1i} x_{1j} + \omega_i^2 m_2 x_{2i} x_{2j} + \ldots + \omega_i^2 m_n x_{ni} x_{nj}$$
$$= \omega_j^2 m_1 x_{1j} x_{1i} + \omega_j^2 m_2 x_{2j} x_{2i} + \ldots + \omega_j^2 m_n x_{nj} x_{ni}$$

which becomes, after rearranging in matrix notation,

$$\omega_i^2 \mathbf{X}_i^T \, \mathbf{M} \mathbf{X}_j = \omega_j^2 \mathbf{X}_j^T \, \mathbf{M} \mathbf{X}_i \tag{4.18}$$

Since the frequencies of the modes are not the same, i.e. $\omega_1 \neq \omega_j$, then

$$\mathbf{X}_i^T \mathbf{M} \mathbf{X}_j = 0 \tag{4.19}$$

and the vectors $\mathbf{X}_i$ and $\mathbf{X}_j$ may be transposed.

A further orthogonality condition can be obtained from Equation (4.18) by replacing the inertia force by the elastic force to give

$$\mathbf{X}_i^T \mathbf{K} \mathbf{X}_j = 0 \tag{4.20}$$

By introducing the notation of Equation (4.16) into Equations (4.19) and (4.20), the orthogonality condition applies to damping, therefore

Fig. 4.2 Arbitrary orthogonal modes.

$$\phi_i^T M \phi_j = 0 \tag{4.21}$$

$$\phi_i^T K \phi_j = 0 \tag{4.22}$$

It is also assumed in subsequent analysis that the orthogonality condition applies to damping, therefore

$$\phi_i^T C \phi_j = 0 \tag{4.23}$$

Equations (4.21) to (4.23) are valid for any two modes of vibration of a system provided their frequencies are not the same.

4.6   Mode superposition

The formulation of the equations of motion of a MDOF system, in a matrix form, has been presented in the previous section. Since the stiffness matrix, which is square and symmetric, has off-diagonal terms, the equations of motion are coupled. The particular case of free undamped motion has been examined for which the frequencies and the shapes of the modes may be obtained. The shapes of the modes were defined by the displacement vector, X. The determination of the free undamped vibration characteristics of a MDOF system is an essential prerequisite in the analysis of the damped forced response of structures.

The modal superposition method of analysis uses an alternative definition of the displacements. In section 1.5 of Chapter 1, the concept of the deflected shape of a structure being defined by the sum of a series of deflected shapes, was discussed. In the present context, the displaced shape of a structure will be represented by the sum of the free vibration mode shapes. For an nDOF system, there will be n independent displacement patterns, the amplitudes of which are considered to be generalized co-ordinates for the actual displacement. The structure, which is shown in Fig. 4.3, illustrates how the final deflected shape can be represented by the sum of the separate modes. From this figure

$$X = \phi_1 Y_1 + \phi_2 Y_2 + \ldots + \phi_n Y_n$$

which may be written in matrix notation as

$$X = \phi Y \tag{4.24}$$

where ϕ is the mode shape matrix relating to the generalized or normal co-ordinates Y to the previously considered local co-ordinates X. By multiplying Equation (4.24) by $\phi_n^T M$, the normal co-ordinate, Y_n, of an arbitrary mode, n, is obtained as follows

$$\phi_n^T M \phi Y = \phi_n^T M \phi_1 Y_1 + \phi_n^T M \phi_2 Y_2 + \ldots + \phi_n^T M \phi_N Y_N \tag{4.25}$$

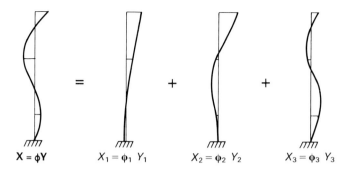

$X = \phi Y$ $X_1 = \phi_1 \, Y_1$ $X_2 = \phi_2 \, Y_2$ $X_3 = \phi_3 \, Y_3$

Fig. 4.3 Deflections as a sum of model components.

Equation (4.25) is considerably simplified by the property of orthogonality, since all terms for which $n \neq N$ are reduced to zero, hence for the nth mode

$$\phi_n^T M X = \phi_n^T M \phi_n Y_n$$

and

$$Y_n = \frac{\phi_n^T M X}{\phi_n^T M \phi_n} \tag{4.26}$$

The generalized co-ordinates may be substituted for the local co-ordinates and their derivatives in Equation (4.10) to give

$$M \phi \ddot{Y} + K \phi Y = 0 \tag{4.27}$$

for undamped free vibration, or

$$M \phi \ddot{Y} + K \phi Y = P(t) \tag{4.28}$$

for undamped forced vibration.

   In order to utilize the simplification of orthogonality of mode shapes, Equation (4.28) is premultiplied by ϕ_n^T, i.e.

$$\phi_n^T M \phi_n \ddot{Y} + \phi_n^T K \phi_n Y = \phi_n^T P(t) \tag{4.29}$$

which becomes (see Equation (4.5))

$$\phi_n^T M \phi_n \ddot{Y}_n + \phi_n^T K \phi_n Y_n = \phi_n^T P(t) \tag{4.30}$$

Equation (4.30) can be rewritten as

$$M_n \ddot{Y}_n + K_n Y_n = P_n(t) \tag{4.31}$$

where M_n, K_n and $P_n(t)$ are the generalized mass, stiffness and loading for mode n, respectively, and are given by

$$M_n = \phi_n^T M \phi_n; \quad K_n = \phi_n^T K \phi_n; \quad P_n(t) = \phi_n^T P(t)$$

Since $\phi_n^T K \phi_n = \omega_n^2 \phi_n^T M \phi_n$, then $K_n = \omega_n^2 M_n$, which is analogous to the SDOF relationship.

Equation (4.28) can be extended to include damping, thus

$$\mathbf{M}\boldsymbol{\phi}\ddot{\mathbf{Y}} + \mathbf{C}\boldsymbol{\phi}\dot{\mathbf{Y}} + \mathbf{K}\boldsymbol{\phi}\mathbf{Y} = \mathbf{P(t)} \tag{4.32}$$

By using Equation (4.23), Equation (4.31) becomes

$$M_n\ddot{Y}_n + C_n\dot{Y}_n + K_nY_n = P_n(t) \tag{4.33}$$

If C_n is assumed to equal $2\xi_n\omega_nM_n$, then Equation (4.23) becomes,

$$\ddot{Y}_n + 2\xi_n\omega_n\dot{Y}_n + \omega_n^2Y_n = P_n(t)/M_n \tag{4.34}$$

It may be seen, therefore, that the coupled equations of motion in local co-ordinates, such as those given in Equation (4.4), may be uncoupled by the use of normal co-ordinates. A SDOF equation is obtained for each mode and the modal superposition method requires that the total response is obtained by the superposition of each modal response.

4.6.1 Damping orthogonality

The orthogonality conditions for mass and stiffness have been demonstrated in section 4.6. In order to satisfy the same conditions for damping, Equation (4.23) was introduced. This equation is satisfied if the damping matrix is assumed to be of the form

$$\mathbf{C} = \lambda_m\mathbf{M} + \lambda_k\mathbf{K} \tag{4.35}$$

The orthogonality relationships given by Equations (4.21) and (4.22) are only two of many such relationships involving the mode shape vector and the mass and stiffness matrices. By successive multiplication it may be shown that a complete family of relationships may be given by

$$\boldsymbol{\phi}_i^T\mathbf{M}[\mathbf{M}^{-1}\mathbf{K}]^r\boldsymbol{\phi}_j = 0 \quad\quad -\infty < r < \infty \tag{4.36}$$

Equations (4.22) and (4.23) are obtained when $r = 0$ and $+1$ respectively. A more general form of damping matrix is, therefore,

$$\mathbf{C} = \mathbf{C}_1 + \mathbf{C}_2 + \ldots + \mathbf{C}_r = \mathbf{M}\sum_r\lambda_r\mathbf{M}^{-1}\mathbf{K}^r \tag{4.37}$$

Since for each mode, C_n equals $\boldsymbol{\phi}_n^T\mathbf{C}\boldsymbol{\phi}_n$ or $2\xi_n\omega_nM_n$, each term in the series of Equation (4.37) is given by

$$C_{nr} = \lambda_r\boldsymbol{\phi}_n^T\mathbf{M}[\mathbf{M}^{-1}\mathbf{K}]^r\boldsymbol{\phi}_n \tag{4.38}$$

after some manipulation

$$C_{nr} = \lambda_r\omega_n^{2r}M_n \tag{4.39}$$

and

$$C_n = 2\xi_n\omega_nM_n = \sum_r\lambda_r\omega_n^{2r}M_n \tag{4.40}$$

from which the damping ratios become

$$\xi_n = \frac{1}{2\omega_n} \sum_r \lambda_r \omega_n^{2r} \tag{4.41}$$

The constants λ_r are obtained from Equation (4.41) if the damping ratios, ξ_n, have been provided for a specified number of modes, hence, the resulting damping matrix. The values of r for the calculation of λ_r are selected to be close to zero.

The following example demonstrates the application of the modal superposition method to the dynamic analysis of structures. It should be noted that the method may be used to evaluate the dynamic response of linear structures for which a local co-ordinate system describes the displacements and the damping is expressed by modal damping ratios.

Example 4.2

The undamped 3DOF system shown in Fig. E4.1 has an initial displaced shape before being released into free vibration. If the displaced shape at $t = 0$ is $x_1 = 6\,\text{mm}$, $x_2 = -16\,\text{mm}$, $x_3 = 6\,\text{mm}$, determine, using the mode superposition method, the displaced shape at $t = 2\pi/\omega_1$.

The solution is obtained by using the following procedure for the mode superposition method.

Step 1   Determine the natural frequencies and associated normalized mode shapes. For structures with many DOF only the first few modes need be considered.

For this example the mode shapes and natural frequencies are, from Example 4.1

$$\phi = \begin{bmatrix} 1 & 1 & 1 \\ 0.547 & -1.49 & -6.05 \\ 0.197 & -0.86 & 11.70 \end{bmatrix} \qquad \omega = \begin{bmatrix} 11.62 \\ 27.50 \\ 45.94 \end{bmatrix} \quad \text{rad/s}$$

Step 2   Calculate the generalized mass, M_n, and generalized load, P_n, using Equation (4.31).

For free vibration, $P_n = 0$ and only M_n has to be determined. The mass matrix, from Example 4.1, is

$$\mathbf{M} = 10^5 \begin{bmatrix} 4 & 0 & 0 \\ 0 & 4 & 0 \\ 0 & 0 & 4 \end{bmatrix} \quad \text{kg}$$

and from Equation (4.31)

$$M_n = \phi_n^T \mathbf{M} \, \phi_n$$

$$M_1 = 10^5 \, [1 \; 0.547 \; 0.197] \begin{bmatrix} 400 \\ 040 \\ 004 \end{bmatrix} \begin{bmatrix} 1 \\ 0.547 \\ 0.197 \end{bmatrix} = 5.361 \times 10^5$$

$$M_2 = 10^5 \, [1 \; -1.49 \; -0.86] \begin{bmatrix} 400 \\ 040 \\ 004 \end{bmatrix} \begin{bmatrix} 1 \\ -1.49 \\ -0.86 \end{bmatrix} = 15.839 \times 10^5$$

$$M_3 = 10^5 \, [1 \; -6.05 \; 11.70] \begin{bmatrix} 400 \\ 040 \\ 004 \end{bmatrix} \begin{bmatrix} 1 \\ -6.05 \\ -11.70 \end{bmatrix} = 697.97 \times 10^5$$

Step 3   Change to a generalized co-ordinate system (Equation (4.24)) and calculate the initial values of displacement, velocity and acceleration using Equation (4.26).

$$Y_n(0) = \frac{\phi_n^T \, \mathbf{M} \, \mathbf{X}(0)}{M_n}; \quad \dot{Y}_n(0) = \frac{\phi_n^T \mathbf{M} \dot{\mathbf{X}}(0)}{M_n}; \quad \ddot{Y}_n(0) = \frac{\phi_n^T \mathbf{M} \ddot{\mathbf{X}}(0)}{M_n}$$

$$Y_1(0) = \frac{[1 \; 0.547 \; 0.197] \begin{bmatrix} 400 \\ 040 \\ 004 \end{bmatrix} \times 10^5 \begin{bmatrix} 6 \\ -16 \\ 6 \end{bmatrix} \times 10^{-3}}{5.361 \times 10^5} = -1.171 \times 10^{-3}$$

$$Y_2(0) = \frac{[1 \; -1.49 \; -0.86] \begin{bmatrix} 400 \\ 040 \\ 004 \end{bmatrix} \begin{bmatrix} 6 \\ -16 \\ 6 \end{bmatrix} \times 10^2}{15.839 \times 10^5} = 6.23 \times 10^{-3}$$

$$Y_3(0) = \frac{[1 \; -6.05 \; 11.70] \begin{bmatrix} 400 \\ 040 \\ 004 \end{bmatrix} \begin{bmatrix} 6 \\ -16 \\ 6 \end{bmatrix} \times 10^2}{697.97 \times 10^5} = 0.991 \times 10^{-3}$$

The initial velocities and accelerations are equal to zero.

Step 4   Solve the uncoupled equations of motion given in Equation (4.34).
   For an undamped structure in free vibration the uncoupled equations of motion are obtained from

$$\ddot{Y}_n + \omega^2 Y_n = 0$$

The solution of this equation, in terms of the initial conditions, is

$$Y_n = \frac{\dot{Y}_n(0)}{\omega} \sin \omega t + Y_n(0) \cos \omega t$$

Substituting the initial values for each mode gives

$$Y_1(t) = -1.171 \times 10^{-3} \cos \omega_1 t$$
$$Y_2(t) = 6.23 \times 10^{-3} \cos \omega_2 t$$

$$Y_3(t) = 0.991 \times 10^{-3} \cos \omega_3 t$$

Step 5 The total response is obtained by summing individual modal responses.

From equation (4.24)

$$\mathbf{X(t)} = \boldsymbol{\phi} \mathbf{Y} = \boldsymbol{\phi}_1 Y_1 + \boldsymbol{\phi}_2 Y_2 + \boldsymbol{\phi}_3 Y_3$$

Hence the total response, in millimetres, is

$$\mathbf{X(t)} = \begin{bmatrix} x_1 \\ x_2 \\ x_3 \end{bmatrix} = - \begin{bmatrix} 1 \\ 0.547 \\ 0.197 \end{bmatrix} 1.171 \cos \omega_1 t + \begin{bmatrix} 1 \\ -1.49 \\ -0.86 \end{bmatrix} 6.23 \cos \omega_2 t$$

$$+ \begin{bmatrix} 1 \\ -6.05 \\ 11.70 \end{bmatrix} 0.991 \cos \omega_3 t \quad \text{mm}$$

At $t = 2\pi/\omega_1$,

$$\mathbf{X}\!\left(\frac{2\pi}{\omega_1}\right) = \begin{bmatrix} x_1 \\ x_2 \\ x_3 \end{bmatrix} = - \begin{bmatrix} 1 \\ 0.547 \\ 0.197 \end{bmatrix} 1.171 - \begin{bmatrix} 1 \\ -1.49 \\ -0.86 \end{bmatrix} 4.167$$

$$+ \begin{bmatrix} 1 \\ -6.05 \\ 11.70 \end{bmatrix} 0.949 \quad \text{mm}$$

The displaced shape at $t = 2\pi/\omega_1$ is

$$\mathbf{X(t)} = \begin{bmatrix} x_1 \\ x_2 \\ x_3 \end{bmatrix} = \begin{bmatrix} -4.39 \\ -0.17 \\ 14.46 \end{bmatrix} \quad \text{mm}$$

4.7 Approximate methods

Approximate methods of dynamic analysis are particularly attractive in practical engineering situations. This is because the information regarding the dynamic properties of a system and the loads to which the system is subjected may not be known exactly. In such cases it would seem inappropriate, therefore, to employ exact methods of analysis when the input information to a problem is known only approximately.

It has been shown how the mode superposition method enables the dynamic response of a structural system to be obtained from the vibration mode shapes and frequencies. In many engineering situations it is often sufficient to obtain the response of only a few modes and these define the response sufficiently accurately for engineering purposes.

In this chapter, two approximate methods are proposed which may be used in a design office with the aid of a calculator. The dynamic analysis of structures with up to ten degrees of freedom may be undertaken conveniently using these methods.

Both the Stodola method and the Holzer method are approximate techniques which are based upon matrix iteration. No formal programming has been provided, but the matrix operations involved in the methods are standard procedures in matrix algebra and may be easily carried out using a computer. The Stodola method requires an initial assumption to be made for the vibration mode shape which is then adjusted iteratively to an acceptable degree of accuracy. The vibration frequency is subsequently obtained. The Holzer method operates on an assumed vibration frequency which is adjusted iteratively to an acceptable degree of accuracy. The vibration frequency is subsequently obtained. The Holzer method operates on an assumed vibration frequency which is adjusted iteratively to satisfy the boundary conditions. The mode shape is then automatically obtained.

4.7.1   The Stodola method

Equation (4.14) may be written as

$$\frac{1}{\omega^2}\mathbf{X} = \mathbf{F}\mathbf{M}\mathbf{X} \tag{4.42}$$

or

$$\frac{1}{\omega^2}\mathbf{X} = \mathbf{H}\mathbf{X} \tag{4.43}$$

where $\mathbf{H} = \mathbf{F}\mathbf{M}$.

Equation (4.43) is only satisfied if the vector $\mathbf{X}$, which represents the mode shapes, is correct. There will be N shapes for the N modes of vibration governed by Equation (4.43).

The Stodola method proceeds initially by assuming a first mode shape which is substituted into Equation (4.43). The first mode shape is assumed to be given by

$$\mathbf{X}_1^0 \quad \begin{array}{l} - \text{ initial assumption} \\ \\ - \text{ first mode} \end{array}$$

and is pre-multiplied by $\mathbf{H}_1$ to give a new shape, i.e.

$$\frac{1}{\omega_1^2}\mathbf{X}_1^1 = \mathbf{H}_1\mathbf{X}_1^0 \tag{4.44}$$

Since $1/\omega_1^2$ is unknown, the left-hand side of Equation (4.44) is rewritten as

$$\overline{\mathbf{X}}_1^1 = \frac{1}{\omega_1^2}\mathbf{X}_1^1 \tag{4.45}$$

The new shape will be different from the initially assumed shape unless the initial shape was correct. For convenience, the values in $\overline{\mathbf{X}}_1^1$ are normalized by dividing the components in the vector by the largest value of displacement. This normalized vector is premultiplied by $\mathbf{H}$ to give a new left-hand side to Equation (4.45) which is designated $\overline{\mathbf{X}}_1^2$. The vector is again normalized and the process is repeated until covergence.

The corresponding frequency of vibration may be determined if the shape of the first mode of vibration has been obtained with sufficient accuracy after say p cycles of iteration. Since further cycles of iteration will not improve the result significantly, the following equation, which is similar to Equation (4.45), may be written

$$\overline{\mathbf{X}}_1^p = \frac{1}{\omega_1^2}\mathbf{X}_1^{p-1} \tag{4.46}$$

and for any point i, Equation (4.46) becomes

$$\bar{x}_{i1}^p = \frac{1}{\omega_1^2}x_{i1}^{p-1} \tag{4.47}$$

$$\omega_1^2 = \frac{x_{i1}^{p-1}}{\bar{x}_{i1}^p} \tag{4.48}$$

If the correct first mode shape is obtained after p cycles of iteration, the frequency of vibration may be obtained from Equation (4.48) using any co-ordinate (usually the position corresponding to the maximum displacement). Less accurate values for the frequency may be obtained by substituting values of x_{i1} and $\bar{x}_{i1}$ which have been obtained prior to the final cycles of iteration.

The following example demonstrates how the Stodola method may be used to obtain the first mode shape and natural frequency.

Example 4.3

Evaluate the first mode shape and natural frequency of the undamped 3DOF system shown in Fig. E4.1 using the Stodola method.

The Stodola method is based upon Equation (4.42)

$$\frac{1}{\omega^2}\mathbf{X} = \mathbf{F}\mathbf{M}\mathbf{X}$$

The solution is based on the following steps.

Step 1 Assume a first mode shape $\mathbf{X}_1^0$ which should be normalized.

$$\text{Initial shape, } \mathbf{X}_1^0 = \begin{bmatrix} 1 \\ 1 \\ 1 \end{bmatrix}$$

Step 2 Premultiply this initial shape by $\mathbf{FM}$ to give a new mode shape (Equation (4.43))

$$\frac{1}{\omega_1^2} \mathbf{X}_1^1 = \mathbf{FMX}_1^0 = \mathbf{HX}_1^0$$

Since ω_1^2 is unknown, the left-hand side of the equation is written as $\bar{\mathbf{X}}_1^1$

$$\bar{\mathbf{X}}_1^1 = \mathbf{HX}_1^0$$

Step 3 $\bar{\mathbf{X}}_1^1$ is normalized by dividing the components in the vector by the largest value.

Step 4 The normalized vector is called $\mathbf{X}_1^1$ and is premultiplied by $\mathbf{H}$ to give a new mode shape $\bar{\mathbf{X}}_1^2$.

$$\bar{\mathbf{X}}_1^2 = \mathbf{HX}_1^1$$

Step 5 The vector is normalized again and this process is repeated until convergence.

$$\bar{\mathbf{X}}_1^p = \mathbf{HX}_1^{p-1} \qquad \text{where } p = \text{number of cycles of iteration}$$

Step 6 The corresponding natural frequency is obtained from Equation (4.48)

$$\omega_1^2 = \frac{x_{i1}^{p-1}}{\bar{x}_{i1}^p}$$

If the correct mode shape is obtained, any co-ordinate in the vector may be used but it is usually the maximum value that is chosen.

For this example the flexibility matrix may be obtained by inverting the stiffness matrix or by applying unit forces to the structure at each floor and determining the corresponding displacements.

$$\mathbf{F} = \frac{1}{480 \times 10^6} \begin{bmatrix} 7 & 3 & 1 \\ 3 & 3 & 1 \\ 1 & 1 & 1 \end{bmatrix}$$

$$\mathbf{H} = \mathbf{FM} = \frac{1}{480 \times 10^6} \begin{bmatrix} 7 & 3 & 1 \\ 3 & 3 & 1 \\ 1 & 1 & 1 \end{bmatrix} \begin{bmatrix} 4 & 0 & 0 \\ 0 & 4 & 0 \\ 0 & 0 & 4 \end{bmatrix} \times 10^5 = \frac{1}{1\,200} \begin{bmatrix} 7 & 3 & 1 \\ 3 & 3 & 1 \\ 1 & 1 & 1 \end{bmatrix}$$

The following table contains the iterative procedure starting with an initial assumption X_1^0. Note that, for ease of calculation, the factor 1 200 has been omitted.

X_1^0	$\bar{X}_1^1$	X_1^1	$\bar{X}_1^2$	X_1^2	$\bar{X}_1^3$	X_1^3	$\bar{X}_1^4$	X_1^4	$\bar{X}_1^5$	X_1^5
1	11	1	9.181	1	8.90	1	8.852	1	8.842	1
1	7	0.636	5.181	0.564	4.90	0.551	4.852	0.548	4.842	0.548
1	3	0.273	1.909	0.208	1.772	0.199	1.750	0.198	1.746	0.197

The mode shape has converged sufficiently to

$$\phi_1 = \begin{bmatrix} 1 \\ 0.548 \\ 0.197 \end{bmatrix}$$

and the frequency is calculated using step 6

$$\omega_1^2 = \frac{x_{11}^4}{\bar{x}_{11}^5} = \frac{1}{\frac{1}{1200} \times 8.842} = 135.7$$

$$\omega_1 = 11.65 \, \text{rad/s}$$

The Stodola iteration procedure will now be developed to obtain the second vibration mode shape. The initially assumed shape of the first mode is substituted into Equation (4.24) which is then expanded

$$X_1^0 = \phi Y^0 = \phi_1 Y_1^0 + \phi_2 Y_2^0 + \ldots + \phi_n Y_n^0 \tag{4.49}$$

By multiplying both sides of Equation (4.49) by $\omega_1^2 M$, an expression is obtained for the inertia forces associated with the assumed initial first mode shape. A further multiplication of Equation (4.49) by the flexibility matrix F gives the second cycle of deflections for the first mode resulting from the inertia forces. After some rearranging, these deflections may be written as

$$X_1^1 = F M \left[\phi_1 \omega_1^2 Y_1^0 + \phi_2 \omega_2^2 Y_2^0 \left(\frac{\omega_1}{\omega_2} \right) + \ldots + \phi_n \omega_n^2 Y_n^0 \left(\frac{\omega_1}{\omega_n} \right) \right] \tag{4.50}$$

Since $F M \phi_n \omega_n^2 = \phi_n$ (from $M \omega^2 = K$), Equation (4.50) may also be written as

$$X_1^1 = \phi_1 Y_1^0 + \phi_2 Y_2^0 \left(\frac{\omega_1}{\omega_2} \right)^2 + \ldots + \phi_n Y_n^0 \left(\frac{\omega_1}{\omega_n} \right)^2 \tag{4.51}$$

By repeating the above procedure, the deflected shape corresponding to the second cycle of iteration is given by

$$\mathbf{X}_1^2 = \boldsymbol{\phi}_1 Y_1^0 + \boldsymbol{\phi}_2 Y_2^0 \left(\frac{\omega_1}{\omega_2}\right)^4 + \ldots + \boldsymbol{\phi}_n Y_n^0 \left(\frac{\omega_1}{\omega_n}\right)^4 \tag{4.52}$$

If the procedure is continued, the following expression is obtained after p cycles of iteration

$$\mathbf{X}_1^p = \boldsymbol{\phi}_1 Y_1^0 + \boldsymbol{\phi}_2 Y_2^0 \left(\frac{\omega_1}{\omega_2}\right)^{2p} + \ldots + \boldsymbol{\phi}_n Y_n^0 \left(\frac{\omega_1}{\omega_n}\right)^{2p} \tag{4.53}$$

which may be generalized for any mode shape by omitting some of the subscripts as follows

$$\mathbf{X}^p = \boldsymbol{\phi}_1 Y_1^0 \left(\frac{\omega}{\omega_1}\right)^{2p} + \boldsymbol{\phi}_2 Y_2^0 \left(\frac{\omega}{\omega_2}\right)^{2p} + \ldots + \boldsymbol{\phi}_n Y_n^0 \left(\frac{\omega}{\omega_n}\right)^{2p} \tag{4.54}$$

If Y_1^0 is zero, the expression converges to the second mode shape and if Y_1^0 and Y_2^0 are both zero, the expression converges to the third mode shape, and so on.

The procedure to obtain the second mode shape is given in what follows. An equation, similar to Equation (4.49), is written for the assumed second mode shape, i.e.

$$\mathbf{X}_2^0 = \boldsymbol{\phi} \mathbf{Y}^0 \tag{4.55}$$

Multiplying both sides by $\boldsymbol{\phi}_1^T \mathbf{M}$ and expanding

$$\boldsymbol{\phi}_1^T \mathbf{M} \mathbf{X}_2^0 = \boldsymbol{\phi}_1^T \mathbf{M} \boldsymbol{\phi}_1 Y_1^0 + \boldsymbol{\phi}_1^T \mathbf{M} \boldsymbol{\phi}_2 Y_2^0 + \ldots + \boldsymbol{\phi}_1^T \mathbf{M} \boldsymbol{\phi}_n Y_n^0 \tag{4.56}$$

Only the first term on the right-hand side is retained due to the orthogonality condition, therefore,

$$Y = \frac{\boldsymbol{\phi}_1^T \mathbf{M} \mathbf{X}_2^0}{\boldsymbol{\phi}_1^T \mathbf{M} \boldsymbol{\phi}_1} \tag{4.57}$$

If the contribution of the first mode is eliminated from Equation (4.55), the remaining terms will converge towards the second mode shape. The first mode contribution has been seen to be given by $\boldsymbol{\phi}_1 Y_1^0$ and this term is subtracted from the right-hand side of Equation (4.55) to give a shape which converges towards the second mode shape

$$\hat{\mathbf{X}}_2^0 = \boldsymbol{\phi} \mathbf{Y}^0 - \boldsymbol{\phi}_1 \frac{\boldsymbol{\phi}_1^T \mathbf{M} \mathbf{X}_2^0}{\boldsymbol{\phi}_1^T \mathbf{M} \boldsymbol{\phi}_1} \tag{4.58}$$

$$\hat{\mathbf{X}}_2^0 = \left[\mathbf{I} - \boldsymbol{\phi}_1 \frac{\boldsymbol{\phi}_1^T \mathbf{M}}{\boldsymbol{\phi}_1^T \mathbf{M} \boldsymbol{\phi}_1}\right] \mathbf{X}_2^0 \tag{4.59}$$

which may be written as

$$\hat{\mathbf{X}}_2^0 = \mathbf{Q}_1 \mathbf{X}_2^0 \tag{4.60}$$

where

$$Q_1 = I - \phi_1 \frac{\phi_1^T M}{\phi_1^T M \phi_1}$$

and I is the identity matrix.

Equation (4.44) is rewritten for the second mode shape with the vectors $\hat{X}_2^0$ and $\hat{X}_2^1$ having no contribution for the first mode, i.e.

$$\frac{1}{\omega_2^2} \hat{X}_2^1 = H_1 \hat{X}_2^0 \tag{4.61}$$

Substituting Equation (4.60) into Equation (4.61)

$$\frac{1}{\omega_2^2} \hat{X}_2^1 = H_1 Q_1 X_2^0 = H_2 X_2^0 \tag{4.62}$$

The matrix H_2 eliminates the contribution of the first mode from the assumed initial shape of the second mode X_2^0 which converges to the true shape. The subsequent iterative procedure then follows that for the first mode shape. The frequency of vibration corresponding to the second mode is obtained from Equation (4.48) with all the terms relating to the second mode.

If the contribution of the first mode $\phi_1 Y_2$ and the second mode $\phi_2 Y_2$ are eliminated from the initially assumed mode shape, the resultant shape vector will converge towards the third mode shape.

An assumed initial third mode shape may be

$$\hat{X}_3^0 = X_3^0 - \phi_1 Y_1^0 - \phi_2 Y_2^0 \tag{4.63}$$

By multiplying Equation (4.63) by $\phi_1^T M$ and then by $\phi_2^T M$ and applying the condition that $\hat{X}_3^0$ be orthogonal to both ϕ_1 and ϕ_2 such that $\phi_1^T M \hat{X}_3^0 = \phi_2^T M \hat{X}_3^0 = 0$, the following two equations are obtained

$$0 = \phi_1^T M X_3^0 - \phi_1^T M \phi_1 Y_1^0 \tag{4.64}$$

$$0 = \phi_2^T M X_3^0 - \phi_2^T M \phi_2 Y_2^0 \tag{4.65}$$

from which

$$Y_1^0 = \frac{\phi_1^T M X_3^0}{\phi_1^T M \phi_1} \tag{4.66}$$

$$Y_2^0 = \frac{\phi_2^T M X_3^0}{\phi_2^T M \phi_2} \tag{4.67}$$

and these equations are equivalent to Equation (4.57).

Substitution of Equations (4.66) and (4.67) into Equation (4.63) gives, after some rearranging,

$$\hat{X}_3^0 = \left[I - \phi_1 \frac{\phi_1^T M}{\phi_1^T M \phi_1} - \phi_2 \frac{\phi_2^T M}{\phi_2^T M \phi_2} \right] X_3^0 \tag{4.68}$$

which is equivalent to Equation (4.59). Equation (4.68) may be rewritten as

$$\hat{X}_3^0 = Q_2 X_3^0 \tag{4.69}$$

where

$$Q_2 = Q_1 - \phi_2 \frac{\phi_2^T M}{\phi_2^T M \phi_2} \tag{4.70}$$

Following Equations (4.44), (4.61) and (4.62), the iterative procedure is expressed by the next equation in which the vectors $\hat{X}_3^1$ and $\hat{X}_3^0$ have no first and second mode contributions

$$\frac{1}{\omega_3^2} \hat{X}_3^1 = H_1 \hat{X}_3^0 \tag{4.71}$$

Substituting Equation (4.69) into Equation (4.71) gives

$$\frac{1}{\omega_3^2} \hat{X}_3^1 = H_1 Q_2 X_3^0 = H_3 X_3^0 \tag{4.72}$$

The matrix H_3 eliminates the contributions of the first and second mode components from the initially assumed mode shape which then converges towards the third mode shape.

It should be obvious that general expressions may be written for the important matrices enabling the shape of any mode to be obtained using the iteration procedure, i.e.

$$Q_n = Q_{n-1} - \phi_n \frac{\phi_n^T M}{\phi_n^T M \phi_n} \tag{4.73}$$

and

$$H_{n+1} = H_1 Q_n \tag{4.74}$$

Example 4.4

In Example 4.3 the first mode of the structure shown in Fig. E4.1 was evaluated using the Stodola method. In this example the second mode shape and natural frequency are calculated.

The second mode of vibration is obtained by eliminating the contribution of the first mode (Equations (4.58) to (4.62)). The procedure is then exactly the same as for the first mode. The following steps illustrate the procedure.

Step 1 Calculate the 'sweeping' matrix Q (Equation (4.60)). For higher modes Equation (4.73) is used.

For the second mode

$$Q_1 = I - \frac{1}{M_1} \phi_1 \phi_1^T M$$

where the generalized mass $M_1 = \phi_1^T M \phi_1$ and I is the identity matrix.

$$Q_1 = I - \frac{10^5}{5.361 \times 10^5} \begin{bmatrix} 1 \\ 0.548 \\ 0.197 \end{bmatrix} [1 \ 0.548 \ 0.197] \begin{bmatrix} 400 \\ 040 \\ 004 \end{bmatrix}$$

$$Q_1 = \frac{1}{1.34} \begin{bmatrix} 0.34 & -0.548 & -0.197 \\ -0.548 & 1.040 & -0.108 \\ -0.197 & -0.108 & 1.301 \end{bmatrix}$$

Step 2   Calculate the updated matrix H (Equation (4.62)). For higher modes Equation (4.74) is used.

$$H_2 = H_1 Q_1 = \frac{1}{1\,200} \begin{bmatrix} 731 \\ 331 \\ 111 \end{bmatrix} \frac{1}{1.34} \begin{bmatrix} 0.34 & -0.548 & -0.197 \\ -0.548 & 1.040 & -0.108 \\ -0.197 & -0.108 & 1.301 \end{bmatrix}$$

$$H_2 = \frac{1}{1\,608} \begin{bmatrix} 0.539 & -0.824 & -0.402 \\ -0.821 & 1.368 & 0.386 \\ -0.405 & -0.384 & 0.996 \end{bmatrix}$$

Step 3   Follow steps 1 to 6 in Example 4.3 to determine the second mode shape and frequency.

Take the initial shape as

$$X_2^0 = \begin{bmatrix} 1 \\ -1 \\ -1 \end{bmatrix} \qquad \bar{X}_2^1 = H_2 X_2^0$$

The following table contains the iterative procedure starting with an initial assumption X_2^0.

X_2^0	$\bar{X}_2^1$	X_2^1	$\bar{X}_2^2$	X_2^2	$\bar{X}_2^3$	X_2^3	$\bar{X}_2^4$	X_2^4	$\bar{X}_2^5$	X_2^5
1	1.765	1	2.148	1	2.138	1	2.135	1	2.133	1
-1	-2.575	-1.459	-3.207	-1.493	-3.218	-1.505	-3.221	-1.509	-3.222	-1.511
-1	-1.785	-1.011	-1.972	-0.918	-1.893	-0.885	-1.864	-0.873	-1.854	-0.869

The mode shape has converged sufficiently to

$$\phi_2 = \begin{bmatrix} 1 \\ -1.511 \\ -0.869 \end{bmatrix}$$

and the frequency is

$$\omega_2^2 = \frac{x_{21}^4}{x_{21}^5} = \frac{1}{\dfrac{1}{1\,608} \times 2.133} = 753.9$$

$$\omega_2 = 27.46\,\text{rad/s}$$

4.7.2 The Holzer method

The Holzer method which obtains the free vibration characteristics of a system proceeds in the reverse direction to the Stodola method. An initial frequency of vibration of a mode is assumed which is then successively corrected until the true value is obtained. The shape of the corresponding mode of vibration is simultaneously calculated as the frequency is corrected. The Holzer method is most usefully applied to structures which have their properties defined along a single axis, such as shear frame buildings.

The shear building illustrated in Fig. 4.4 will be used to explain the method. The frame is constrained to move in the horizontal direction with no joint rotations. The mass of each storey is concentrated at the rigid floor slabs and the storey stiffness is associated with the lateral flexure of the columns. The conditions governing motion are, in fact, the same as those for the frame of Fig. 1.3.

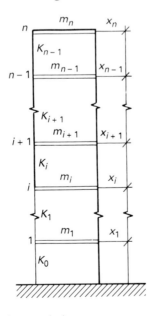

Fig. 4.4 Shear building for Holzer analysis.

The stiffness of the ith storey is given by

$$K = 12 \frac{EI}{l^3} \tag{4.75}$$

where EI is the flexural rigidity of all the columns in a storey and l is the storey height.

Since only free undamped vibrations are being considered, the inertia force at the level corresponding to floor slab i is equal to $\omega^2 m_i x_i$, where ω is the current value of the assumed frequency, m_i is the mass concentrated at the ith level and x_i is the displacement of the ith storey. From a consideration of equilibrium, the shear force S_i in storey i will equal the sum of the inertia forces from floor slab i to floor slab n.

By making an assumption regarding the frequency of vibration and of the amplitude of motion for one end of the structure, it is possible to conduct a series of calculations to determine the forces and displacements for each storey from level n to the base. For instance, a value of unity would be taken as the displacement of level n and a corresponding value of frequency estimated. A series of calculations is then carried out for each storey in turn to determine forces and displacements. The last calculation should satisfy the boundary condition of zero base displacement. Unless the assumed frequency is correct, the boundary condition will not be satisfied and a residual value of displacement will occur. A further estimate of the frequency is made and the procedure repeated until the boundary condition is satisfied. It is usual to interpolate between trial values in order to reduce the number of frequency estimates. A plot is made of base displacement versus estimated frequency from which it is possible to obtain the frequencies of the modes of vibration participating in the motion. Mode frequencies will correspond to zero base displacement.

The Holzer method is probably best illustrated by an example.

Example 4.5

Evaluate the third mode shape and natural frequency of the undamped 3 DOF system shown in Fig. E4.1, using the Holzer method.

The following steps illustrate the Holzer procedure.

Step 1   Assume a value for the frequency, ω. The Holzer method can be used to obtain any of the modes and so a reasonable value for the third mode would be somewhat higher than the second natural frequency. Take $\omega_3 = 40$, i.e. $\omega_3^2 = 1\,600$.

Step 2   The displacement of M_1 is taken as unity (i.e. $X_1 = 1$). The displacement of the next mass in the chain, M_2, is calculated from

$$X_2 = X_1 - \omega^2 M_1 X_1 / k_2$$

For the nth mass, the displacement is

$$X_n = X_{n-1} - \frac{\omega^2}{k_n} \sum_{r=1}^{n-1} M_r X_r$$

Step 3 The last calculation should satisfy the boundary conditions. For a structure fixed at its base, $X_{\text{base}} = 0$. If the boundary conditions are not satisfied then the value of ω is changed and step 2 is carried out again. Once two trial values of ω have been used, it is usual to interpolate or extrapolate to reduce the number of frequency estimates. A plot of base displacement versus frequency is made and the mode frequency corresponding to zero base displacement is obtained.

It is convenient to present the Holzer calculations using a table.

Trial 1.   Take $\omega = 40$ ($\omega^2 = 1\,600$).

Level	X	$M_r X_r$	$\sum M_r X_r$	$\frac{\omega^2}{k_n} \sum M_r X_r$
1	1	4×10^5	4×10^5	5.33
2	−4.33	-17.33×10^5	-13.33×10^5	−8.89
3	4.56	18.23×10^5	4.90×10^5	1.63
4	2.93			

The base displacement is not zero and another value of ω is needed. The mode shape shows that the actual natural frequency is higher than 40 rad/s.

Trial 2.   Take $\omega = 50$ ($\omega^2 = 2\,500$).

Level	X	$M_r X_r$	$\sum M_r X_r$	$\frac{\omega^2}{k_n} \sum M_r X_r$
1	1	4×10^5	4×10^5	8.33
2	− 7.33	-29.33×10^5	-25.33×10^5	−26.39
3	19.56	76.23×10^5	50.89×10^5	26.51
4	− 7.45			

The base displacement is not zero and another value of ω is needed. The mode shape shows that the actual natural frequency is lower than 50 rad/s. The next trial value for ω can now be obtained by linear interpolation between the first two trial values with associated base displacements of 2.93 mm and −7.45 mm.

The square of the frequency, ω^2, for zero base displacement is given by

$$\frac{\omega^2 - 1\,600}{2.93} = \frac{2\,500 - \omega^2}{7.45}$$

$$\omega^2 = 1854$$

Take $\omega^2 = 1\,900$ to ease calculation.

Trial 3.   Take $\omega^2 = 1\,900$.

Level	X	$M_r X_r$	$\Sigma M_r X_r$	$\dfrac{\omega^2}{k_n} \Sigma M_r X_r$
1	1	4×10^5	4×10^5	6.33
2	-5.33	-21.33×10^5	-17.33×10^5	-13.89
3	8.39	33.56×10^5	16.22×10^5	6.42
4	1.97			

Further interpolation is necessary since the relationship between base displacement and ω^2 is not truly linear.

Linear interpolation between $\omega^2 = 1\,900$ and $2\,500$ gives, for zero base displacement, $\omega^2 = 2\,025$.

Trial 4.   Take $\omega^2 = 2\,100$.

Level	X	$M_r X_r$	$\Sigma M_r X_r$	$\dfrac{\omega^2}{k_n} \Sigma M_r X_r$
1	1	4×10^5	4×10^5	7.033
2	-6.03	-24.13×10^5	-20.13×10^5	-17.70
3	11.67	46.68×10^5	26.55×10^5	11.67
4	0			

This satisfies the boundary conditions and the third natural frequency is given by

$$\omega_3 = \sqrt{(2\,110)} = 45.9\,\text{rad/s}$$

and the mode shape by

$$\Phi_3 = \begin{bmatrix} 1 \\ -6.03 \\ 11.67 \end{bmatrix}$$

The Holzer method may be interpreted in a more formal manner than in the previous description. The more formal approach involves a transfer matrix procedure and is, therefore, suited to an automatic calculation. The shear forces and displacements at a particular level in the shear building of Fig. 4.4 may be expressed in terms of the shear forces and displacements of an adjacent level. It is possible to extend the procedure to include levels which are several storeys apart. Thus, through a series of matrix transfer operations, the forces and displacements at the base of the shear building

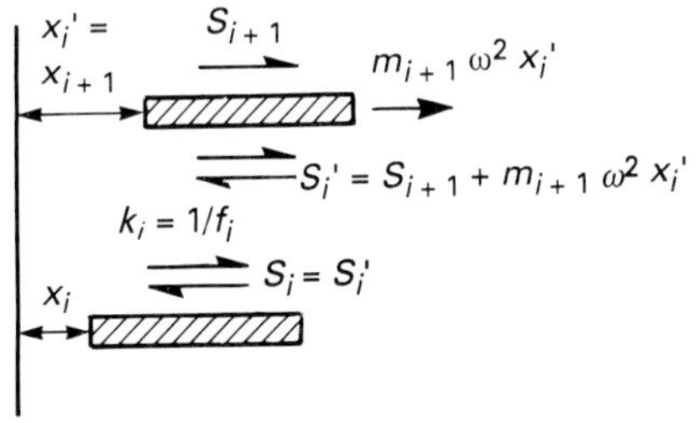

Fig. 4.5 Modified Holzer analysis.

may be related to those of the topmost storey. Figure 4.5 will be used to illustrate the method.

Figure 4.5 shows the relevant shear forces and displacements for the *i*th and $(i + 1)$th levels and the connecting storey. The notation used is the same as before. The dynamic equilibrium and compatibility relationships for these levels and the connecting storey are:

$$S_i = S_i'$$ (4.76)

and

$$x_i = -S_i' \frac{1}{k_i} + x_i'$$ (4.77)

Equations (4.76) and (4.77) can be written in matrix notation as

$$\begin{Bmatrix} S_i \\ x_i \end{Bmatrix} = \begin{bmatrix} 1 & 0 \\ \dfrac{-1}{k_i} & 1 \end{bmatrix} \begin{Bmatrix} S_i' \\ x_i' \end{Bmatrix}$$ (4.78)

Now

$$S_i' = S_{i+1} + \omega^2 m_{i+1} x_{i+1}$$ (4.79)

$$x_i' = x_{i+1}$$ (4.80)

or

$$\begin{Bmatrix} S_i' \\ x_i' \end{Bmatrix} = \begin{bmatrix} 1 & \omega^2 m_{i+1} \\ 0 & 1 \end{bmatrix} \begin{Bmatrix} S_{i+1} \\ x_{i+1} \end{Bmatrix}$$ (4.81)

Equation (4.81) may be combined with Equation (4.78) to give

$$\begin{Bmatrix} S_i \\ x_i \end{Bmatrix} = \begin{bmatrix} 1 & \omega^2 m_{i+1} \\ -k_i^{-1} & 1 - \omega^2 k_i^{-1} m_{i+1} \end{bmatrix} \begin{Bmatrix} S_{i+1} \\ x_{i+1} \end{Bmatrix}$$ (4.82)

or

$$Sx_i = T_{i+1} Sx_{i+1}$$ (4.83)

Thus, the forces and displacements at level i may be expressed in terms of the forces and displacements at level $i + 1$.
    Similarly

$$Sx_{i+1} = T_{i+2}Sx_{i+2} \tag{4.84}$$

The relationship between the forces and displacements at the base or level 0 and the top or level n of the shear building in Fig. 4.4 may be expressed as

$$Sx_0 = T_1T_2T_3 \ldots T_{n-1}T_nSx_n \tag{4.85}$$

or

$$Sx_0 = TSx_n \tag{4.86}$$

where $T = T_1T_2 \ldots T_{n-1}T_n$ and is a 2×2 matrix. Substitution of the boundary conditions $x_0 = 0$ and $S_n = 0$ and of the amplitude assumption $x_n = 1$ enables the values of ω^2 to be obtained from Equation (4.86).

4.8   Distributed systems

4.8.1   Introduction

A variety of structural elements which are of interest to the engineer may be classified as distributed systems for the purpose of dynamic analysis. These elements include beams, plates and structures for which the important structural properties are continuously distributed. In principle, motion of distributed systems is defined by an infinite number of co-ordinates and hence, associated degrees of freedom.
    The equation of motion for these elements may be obtained by deriving an appropriate partial differential equation which is solved when the relevant boundary conditions are applied. In all but the simplest of cases, however, this exact method of solution is extremely tedious and approximate methods are to be preferred.
    In the first part of this section, only elastic beams are considered, for which the effects of rotatory inertia and shear deformation are neglected. Damping effects are also ignored. The partial differential equation of motion is established for these beams and solved for some fundamental cases.
    Section 4.8.5 is more important since it is devoted to the more practical methods of approximate analysis. The Rayleigh and Rayleigh–Ritz energy based methods are introduced and their versatility demonstrated. These energy methods have important implications beyond the scope of structural dynamics.

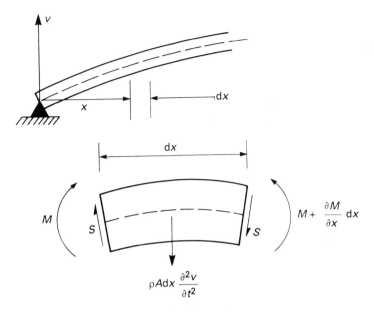

Fig. 4.6   Moments and forces on a vibrating beam element.

Of particular interest to foundation engineers is the response of elastic bars and piles to longitudinal waves. Included, therefore, is a section devoted to this subject and the analysis of stresses induced by piling (section 4.8.6).

4.8.2   Equation of motion

Note that u and v are the longitudinal and transverse beam displacements which are functions of position and time. Taking moments about the centreline (Fig. 4.6)

$$S\,dx + M - \left(M + \frac{\partial M}{\partial x}\,dx\right) = 0 \tag{4.87}$$

where the shear force

$$S = \frac{\partial M}{\partial x} \tag{4.88}$$

for vertical equilibrium, where ρ is the density

$$\frac{\partial S}{\partial x} = \ell \frac{A \partial^2 v}{\partial t^2} \tag{4.89}$$

The incorporation of Equations (4.88) and (4.89) into the differential equation of flexure for the beam

$$M = EI \frac{\partial^2 v}{\partial x^2}$$

produces the following equation of motion in which EI is the beam flexural rigidity which is assumed constant

$$EI \frac{\partial^4 v}{\partial x^4} + \rho \frac{A \partial^2 v}{\partial t^2} = 0 \tag{4.90}$$

The displacement v which is a function of both position and time will be harmonic for the case of free vibrations and may be written as

$$v(x, t) = V(x) \sin(\omega t + \alpha) \tag{4.91}$$

Substituting Equation (4.91) into Equation (4.90) gives

$$\frac{\partial^4 V}{dx^4} - \rho \frac{A \omega^2 V}{EI} = 0 \tag{4.92}$$

Equation (4.92) is a fourth order equation and its general solution is given by

$$V = A_1 \sin \lambda x + A_2 \cos \lambda x + A_3 \sinh \lambda x + A_4 \cosh \lambda x \tag{4.93}$$

in which $\lambda^4 = \rho A \omega^2 / EI$.

The four constants in Equation (4.93) define the shape and amplitude of the beam during its motion and are evaluated by considering the boundary conditions at the ends of the beams. The following two examples will illustrate the procedure of analysis for a simple supported beam and a cantilever beam.

4.8.3 Simply supported beam

For the simply supported beam in Fig. 4.7, the displacements and moments at both supports are zero, i.e.

$$V = 0 \qquad \text{at } x = 0 \text{ and } x = \ell$$

and

$$\frac{\partial^2 v}{\partial x^2} = 0 \qquad \text{at } x = 0 \text{ and } x = \ell$$

Substitution of the boundary conditions at $x = 0$ into Equation (4.93) gives $A_2 = A_4 = 0$. Applying the boundary conditions at $x = \ell$ and eliminating A_2 and A_4 yields

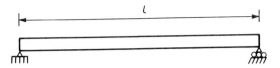

Fig. 4.7 Simply supported beam.

$$0 = -\lambda^2 A_1 \sin \lambda \ell + \lambda^2 A_3 \sinh \lambda \ell \tag{4.94}$$

Since the hyperbolic sine function cannot vanish the only non-trivial solution is $A_3 = 0$ and $\sin \lambda \ell = 0$, therefore

$$\lambda \ell = n\pi \qquad n = 1, 2, 3, \ldots$$

$$\omega_n^2 = \left(\frac{n\pi}{\ell}\right)^4 \frac{EI}{\rho A} \qquad n = 1, 2, 3, \ldots \tag{4.95}$$

The equation $\sin \lambda \ell = 0$ is called the frequency equation for the *n*th mode. Equation (4.95) may be rewritten to define the modes of vibration given by

$$v = A_1 \sin \frac{n\pi x}{\ell} \sin (\omega_n t + \alpha) \tag{4.96}$$

in which the natural frequency of each mode is $\omega_n / 2\pi$. The shapes of the first three modes of vibration are shown in Fig. 4.8.

Mode 1

$$V(x) = \sin \frac{\pi x}{L} \qquad \omega_1 = \frac{\pi^2}{L^2}\sqrt{\left(\frac{EI}{\rho A}\right)}$$

Mode 2

$$V(x) = \sin \frac{2\pi x}{L} \qquad \omega_2 = \frac{4\pi^2}{L^2}\sqrt{\left(\frac{EI}{\rho A}\right)}$$

Mode 3

$$V(x) = \sin \frac{3\pi x}{L} \qquad \omega_3 = \frac{9\pi^2}{L^2}\sqrt{\left(\frac{EI}{\rho A}\right)}$$

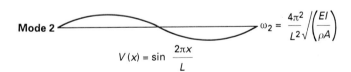

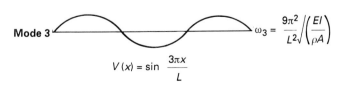

Fig. 4.8 First three modes of vibration of a simply supported beam.

4.8.4 Cantilever beam

The determination of the modes of vibration of a cantilever beam provide a more representative example of the current method of analysis since the coefficients in Equation (4.93), which defines the shape of the beam, do not vanish. With the origin at the left-hand end (Fig. 4.9), the four boundary conditions for the cantilever beam are

$$v = 0 \qquad \text{at } x = 0$$

$$\frac{dv}{dx} = 0 \qquad \text{at } x = 0$$

$$\frac{d^2v}{dx^2} = 0 \qquad \text{at } x = \ell$$

$$\frac{d^3v}{dx^3} = 0 \qquad \text{at } x = \ell$$

Substitution of the boundary conditions into Equation (4.93) yields the following two equations after noting that $A_1 = -A_3$ and $A_2 = -A_4$

$$A_1 \left(-\sin \lambda\ell - \sinh \lambda\ell \right) + A_2 \left(-\cos \lambda\ell - \cosh \lambda\ell \right) = 0 \qquad (4.97)$$

and

$$A_1 \left(-\cos \lambda\ell - \cosh \lambda\ell \right) + A_2 \left(\sin \lambda\ell - \sinh \lambda\ell \right) = 0 \qquad (4.98)$$

which may be written in matrix notation as

$$\begin{bmatrix} \sin \lambda\ell + \sinh \lambda\ell & \cos \lambda\ell + \cosh \lambda\ell \\ \cos \lambda\ell + \cosh \lambda\ell & \sinh \lambda\ell - \sin \lambda\ell \end{bmatrix} \begin{Bmatrix} A_1 \\ A_2 \end{Bmatrix} = \begin{Bmatrix} 0 \\ 0 \end{Bmatrix} \qquad (4.99)$$

Using Cramer's rule to obtain the frequency equation, the determinant of the square matrix is equal to zero if the coefficients are to be non-zero. Obtaining the co-factors and simplifying gives

$$1 + \cos \lambda\ell \cosh \lambda\ell = 0 \qquad (4.100)$$

Equation (4.100) is known as a transcendental equation from which the frequencies of the modes of vibration of the cantilever are obtained. The

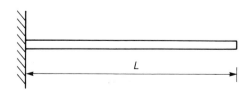

Fig. 4.9 Cantilever beam.

coefficient A_1 may be expressed in terms of A_2 by using Equation (4.97) to give

$$A_2 = -\frac{\sin \lambda \ell + \sinh \lambda \ell}{\cos \lambda \ell + \cosh \lambda \ell} A_1 \qquad (4.101)$$

Continuing Equation (4.101) with $A_1 = -A_3$ and $A_2 = -A_4$ enables Equation (4.93) to be expressed in terms of the coefficient A_1 only, i.e.

$$V = A_1 \left\{ \sin \lambda x - \sinh \lambda x + \left(\frac{\sin \lambda \ell + \sinh \lambda \ell}{\cos \lambda \ell + \cosh \lambda \ell} \right) (\cosh \lambda x - \cos \lambda x) \right\}$$

$$(4.102)$$

The appropriate value of $\lambda \ell$ which is obtained from Equation (4.100) may now be substituted into Equation (4.102) to obtain the corresponding mode shape. The first three modes of vibration and the corresponding frequencies are shown in Fig. 4.10.

4.8.5 Approximate methods

4.8.5.1 Rayleigh's method

Energy is supplied to a vibrating system by the disturbing force and is dissipated by damping effects. During motion, the energy is stored as kinetic energy, T, potential energy, P, and strain energy, V. By applying

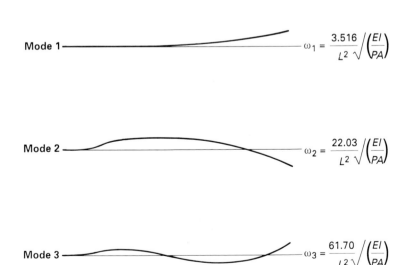

Mode 1 — $\omega_1 = \dfrac{3.516}{L^2} \sqrt{\left(\dfrac{EI}{PA}\right)}$

Mode 2 — $\omega_2 = \dfrac{22.03}{L^2} \sqrt{\left(\dfrac{EI}{PA}\right)}$

Mode 3 — $\omega_3 = \dfrac{61.70}{L^2} \sqrt{\left(\dfrac{EI}{PA}\right)}$

Fig. 4.10 First three modes of vibration of a cantilever beam.

the principle of conservation of energy to an undamped system subjected to free vibration, the following relationship holds

$$V + T + P = \text{a constant} \tag{4.103}$$

The method proposed by Rayleigh may be developed from this statement by particular reference to a single degree of freedom system.

The kinetic energy of the system is

$$T = \frac{1}{2} m \frac{dx(t)}{dt} \tag{4.104}$$

while the strain energy is

$$V = \frac{k}{2} (x(t) + \delta)^2 \tag{4.105}$$

and the potential energy is

$$P = -mg (x(t) + \delta) \tag{4.106}$$

in which $x(t)$ is the displacement from the static position (positive downward) and δ is the static displacement.

Thus for vertical motion

$$\frac{1}{2} m \frac{dx(t)^2}{dt} + \frac{1}{2} k (x(t) + \delta)^2 - mg (x(t) + \delta) = \text{a constant} \tag{4.107}$$

Rearranging and taking constants to the right-hand side

$$\frac{1}{2} m \frac{dx(t)^2}{dt} + \frac{1}{2} k x(t)^2 = \text{a constant} \tag{4.108}$$

Assuming the displacements vary harmonically with the frequency, i.e.

$$x(t) = A \sin \omega t \tag{4.109}$$

and substituting in Equation (4.108), the following relationship is obtained after equating coefficients of $\sin^2 \omega t$

$$\frac{1}{2} m A^2 \omega^2 = \frac{1}{2} k A^2 \tag{4.110}$$

Equation (4.110) states that the maximum kinetic energy equals the maximum strain energy. These energy states occur when the system passes through the equilibrium and extreme positions respectively, i.e.

$$T_{max} = V_{max} \tag{4.111}$$

The relationship given by Equation (4.111) can be applied to the problem of the vibration of a uniform slender beam for which the kinetic and strain energies are given by

$$T_{max} = \frac{1}{2} \omega^2 \int_0^L m\, y(x)^2\, dx$$

$$V_{max} = \frac{1}{2} \int_0^L EI \left[\frac{d^2 y(x)}{dx^2} \right]^2 dx \tag{4.112}$$

where $y(x)$ is the amplitude of the motion or mode shape. Thus, the frequency is obtained from

$$\omega^2 = \frac{\int_0^L EI \left[\frac{d^2 y(x)}{dt^2} \right]^2 dx}{\int_0^L m\, y(x)^2\, dx} \tag{4.113}$$

If the beam has concentrated masses $m_1, m_2, \ldots, m_n$ at $x_1, x_2, \ldots, x_n$, the frequency is obtained from

$$\omega^2 = \frac{\int_0^L EI \left[\frac{d^2 y(x)}{dt^2} \right]^2 dx}{\int_0^L m\, y(x)^2\, dx + \sum_{r=1}^n m_r y(x)^2} \tag{4.114}$$

Equation (4.114) is fundamental to the Rayleigh method in which approximations to the natural frequency are computed from assumptions regarding the mode shape. In practice the method may be used to obtain the frequency of the first mode for which suitable approximations can be made for the shape of simple structural elements. The application of the method will be demonstrated in the following example.

Example 4.6

Determine the natural frequency of the fundamental mode of vibration for the beam shown in Fig. E4.2. The beam has a mass of m per unit length and supports a central point mass M.

  An estimate of the shape of the first mode of vibration is required for the beam. A reasonable choice is that shown in Fig. E4.2 which is the deflected shape of a massless beam subjected to a central point F. The corresponding maximum deflection is assumed to be A. By applying the differential equation of flexure for the beam between A and B, which is

$$EI \frac{d^2 y}{dx^2} = -\frac{Fx}{2}$$

and by integrating and solving for the constants, the deflected shape of the left-hand side of the beam is

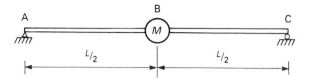

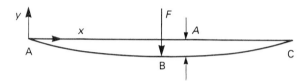

Fig. E4.2

$$y = A\left[3\left(\frac{x}{L}\right) - 4\left(\frac{x}{L}\right)^3\right]$$

The maximum kinetic and strain energies of the system are now required. The maximum kinetic energy of the point mass is given by $\frac{1}{2}M\omega^2A^2$ and the kinetic energy of an element dx of self mass m per unit length is $\frac{1}{2}(m\,dx)\omega^2y^2$. Integrating the latter value for the total beam and remembering that the expression for the deflection is valid between A and B, the total kinetic energy of the whole system is

$$T_{max} = 0.5\,M\omega^2A^2 + 2 \times \frac{1}{2}m\omega^2A^2 \int_0^{L/2} \left[3\left(\frac{x}{L}\right)^3 - 4\left(\frac{x}{L}\right)^3\right]^2 dx$$

$$T_{max} = \omega^2A^2\,(0.5M + 0.243\,mL)$$

Since the strain energy for the assumed deflected shape is $\frac{1}{2}FA$ in which $A = FL^3/48EI$, the maximum strain energy is

$$V_{max} = \frac{24EIA^2}{L^3}$$

Therefore, from Equation (4.111) for the beam, the frequency is obtained from

$$\omega^2 = \frac{24EI}{(0.5M + 0.243\,mL)L^3}$$

An alternative assumption for the deflected shape can be made by considering a beam without a central mass for which case

$$y = \sin\frac{\pi x}{L}$$

This equation may be used instead of the previous expression.

4.8.5.2 Rayleigh—Ritz method

This method may be considered to be an extension of the Rayleigh method and is based upon the concept that a number of functions may be added to provide a more accurate description of a natural mode than might be obtained from the single function used in Rayleigh's method. If suitable functions are chosen then not only are the first mode frequencies estimated but higher mode frequencies may also be obtained. The greater the number of functions chosen the more accurate the result. However, greater computational effort is required

$$y(x) = C_1\phi_1(x) + C_2\phi_2(x) + \ldots + C_n\phi_n(x) \qquad (4.115)$$

The coefficients C_1 to C_n are arbitrary parameters while the functions ϕ_1 to ϕ_n satisfy the boundary conditions for slope and deflection. The arbitrary coefficients are chosen such that the best approximation to the natural modes is obtained by the addition of the terms. This is achieved by selecting the value of the parameter C_n such that the frequencies obtained from Equation (4.114) are a minimum. Equation (4.115) is thus substituted into Equation (4.114) and the resulting expression is partially differentiated with respect to each of the coefficients to give a set of linear homogeneous algebraic equations

$$\frac{\partial\omega^2}{\partial C_1} = 0$$

$$\frac{\partial\omega^2}{\partial C_2} = 0 \qquad (4.116)$$

$$\frac{\partial\omega^2}{\partial C_n} = 0$$

This set of equations represent an eigenvalue problem and may be solved in the usual way. The effectiveness of the Rayleigh—Ritz method depends upon the choice of functions ϕ_1 to ϕ_n.

4.8.6 Longitudinal vibration of elastic bars and piles

The theory of the propagation of stress waves in solids has received considerable attention and the book by Kolsky (1964) provides an introduction to the subject. In this theory, the effects of forces which are applied for only a short time are considered in terms of the propagation of stress waves. Although stress wave propagation in general is an important topic for engineers to consider, the current discussion will be limited to the longitudinal vibration of elastic bars and piles.

Pile driving is a problem in longitudinal wave transmission which may be considered in a general way by the associated wave equation. The structural or foundation engineer is concerned that limiting pile stresses are not exceeded during the pile driving operation. This is particularly so during the driving of precast concrete piles when tensile stresses may be induced.

4.8.6.1 *Equation of longitudinal wave propagation*

Consider the free axial vibration of a bar with cross-sectional area A, modulus of elasticity E, and unit weight γ, (Fig. 4.11). It is assumed that each cross-section remains plane during vibration and that the axial stress is uniform across the area. The stress on a transverse plane at x is σ_x and similarly the stress at $(x + \Delta x)$ is $[\sigma_x + (\partial \sigma_x / \partial x) \Delta x]$. Equilibrium in the x-direction may be written as

$$\left(\sigma_x + \frac{\partial \sigma_x}{\partial x} \Delta x \right) A - \sigma_x A = F \tag{4.117}$$

The force F is provided by the inertial resistance of the element. If u is the displacement of the element in the x-direction, Newton's second law can be written as

$$\left(\sigma_x + \frac{\partial \sigma_x}{\partial x} \Delta x \right) A - \sigma_x A = A \, \Delta x \frac{\gamma}{g} \frac{\partial^2 u}{\partial t^2} \tag{4.118}$$

$$\frac{\partial \sigma_x}{\partial x} = \frac{\gamma}{g} \frac{\partial^2 u}{\partial t^2} \tag{4.119}$$

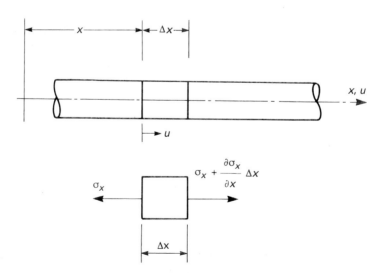

Fig. 4.11 Longitudinal wave propagation in a bar.

The strain in the x-direction is $\partial u/\partial x$ and

$$\sigma_x = E \frac{\partial u}{\partial x} \tag{4.120}$$

Equation (4.119) may therefore be written as

$$E \frac{\partial^2 u}{\partial x^2} = \rho \frac{\partial^2 u}{\partial t^2} \tag{4.121}$$

$$\frac{\partial^2 u}{\partial t^2} = c^2 \frac{\partial^2 u}{\partial x^2} \tag{4.122}$$

$$c^2 = \frac{E}{\rho} \tag{4.123}$$

c is the velocity of longitudinal wave propagation in the bar and $\rho = \gamma/g$ is the mass density. It is easy to verify that Equation (4.122) is satisfied by

$$u = f(x - ct) + g(x + ct) \tag{4.124}$$

where f and g are arbitrary functions of the variables indicated. Equation (4.124) is in fact the general solution of Equation (4.122) and it represents two waves of shapes $u = f(x)$, $u = g(x)$, which do not change, moving respectively in the positive and negative x-directions with velocity c. An alternative description of the solution is to say that the initial values of the waves are propagated along the lines $x - ct = $ constant and $n + ct = $ constant situated in the x, t-plane. These lines are called the characteristics of the equation. This result could be anticipated from a consideration of the consequences of applying an instantaneous displacement, u, to a section of the bar in Fig. 4.11. The cross-section of the bar at $(x + \Delta x)$ would experience a compressive stress and the cross-section at n would experience a tensile stress. With the passage of time, zones further along the bar would experience these stresses caused by the displacement and the result would be a tensile stress wave travelling in the negative x-direction and a compressive wave travelling in the positive x-direction.

From Equations (4.120) and (4.124) the solution in terms of stress becomes

$$\sigma_x = E \frac{\partial u}{\partial x} = E f'(x - ct) + E g'(x + ct) \tag{4.125}$$

The stress wave is also of velocity c and has an unchanging shape.

4.8.6.2 Boundary or end conditions

It has already been stated that the solution of Equation (4.122) is the sum of two arbitrary functions which may be considered to be waves of identical shape and velocity but travelling in opposite directions. Further consideration of these waves will enable the phenomena that occur at the ends of finite bars, through which the waves are travelling, to be understood.

Consider an elastic bar in which a compression wave is travelling in the positive x-direction and an identical tension wave is travelling in the negative y-direction (Fig. 4.12(a)). The stress in the waves is, therefore, $\pm\sigma_x$, and elsewhere in the bar the stress is zero.

During the time interval in which the two waves coincide (Fig. 4.12(b)), the portions of the bar in the two waves are superimposed as zero stress. In fact, the stress on the bar centreline is always zero and is the same as the equilibrium condition at the free end of a bar. By removing one half of the bar, the centreline cross-section can be considered as a free end. From Fig. 4.12(d) it may be seen that a compression wave is reflected from a free end as a tension wave of the same shape and magnitude. Similarly, a tension wave will be reflected as a compression wave.

Now, consider the case of an elastic bar in which compression waves are travelling in both the positive and negative x-directions (Fig. 4.13(a)). During the interval in which the waves traverse the cross-over zones (Fig. 4.13(b)), the centreline stress is twice the compressive stress in each wave. However, throughout the passage of the waves along the bar, the displacement of the centreline cross-section remains zero and, hence, it behaves like the fixed end of a bar. The centreline of the bar may be considered as

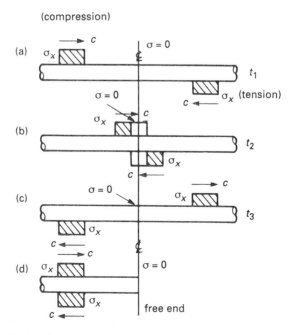

Fig. 4.12   Wave reflection from a free end.

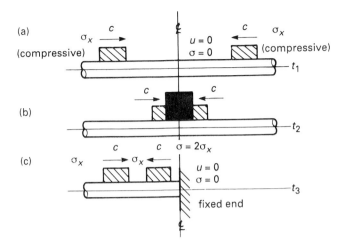

Fig. 4.13   Wave reflection from a fixed end.

a fixed end by the removal of one half of the bar. It can be seen from Fig. 4.13(c) that a compression wave is reflected from the fixed end of a bar as a compression wave of identical shape and magnitude. Also, the stress at the fixed end is twice the compressive wave stress.

Although the incident and reflected waves have been of constant stress, the same conclusions regarding wave propagation and superposition can be applied to stress waves of any shape.

4.8.6.3 *Composite bars*

The establishment of equilibrium and compatibility conditions at a discontinuity within a composite bar demonstrates that reflected and refracted waves are generated as a result of the impingement of an incident wave. The establishment of these conditions will now be considered with the assistance of the previous section.

Figure 4.14 shows the junction between two different sections of a composite bar and the relevant properties of each section. It is assumed that a longitudinal wave propagates in the positive x-direction and impinges on the junction Y–Y.

Since compatibility of displacements must be maintained either side of the junction, the displacement of the refracted wave must equal the sum of the displacements and reflected waves, i.e.

$$u = u_i + u_r \tag{4.126}$$

in which u is the displacement of the refracted wave and u_i and u_r are the incident and reflected wave displacements respectively. The incident and

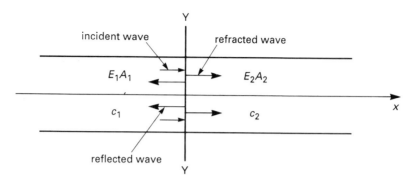

Fig. 4.14 Longitudinal wave behaviour within a composite bar.

reflected waves occur in the bar material to the left of the junction while the refracted wave occurs to the right of the junction.

Similarly, the equilibrium of forces either side of the junction corresponding to displacements may be written as

$$F = F_i + F_r \tag{4.127}$$

Equation (4.126) may be differentiated with respect to time to give

$$\frac{\partial u}{\partial t} = \frac{\partial u_i}{\partial t} + \frac{\partial u_r}{\partial t} \tag{4.128}$$

The incident wave which is propagating in the positive *x*-direction may be represented by the first solution of Equation (4.124), i.e.

$$u_i = f_i(x - ct) \tag{4.129}$$

After some manipulation between Equations (4.128) and (4.129), the time and displacement derivatives are found to be related by

$$\frac{\partial u_i}{\partial t} = -c_1 \frac{\partial u_i}{\partial x} \tag{4.130a}$$

A similar result may be obtained for the reflected and refracted waves, i.e.

$$\frac{\partial u_r}{\partial t} = -c_1 \frac{\partial u_r}{\partial x} \tag{4.130b}$$

and

$$\frac{\partial u}{\partial t} = -c_2 \frac{\partial u}{\partial x} \tag{4.130c}$$

Since terms such as $\partial u_i/\partial x$ represent strain, substitution of Equations (4.130) into Equation (4.128) gives

$$F \frac{c_2}{E_2 A_2} = F_i \frac{c_1}{E_1 A_1} - F_r \frac{c_1}{E_1 A_1} \tag{4.131}$$

Equation (4.131) can be combined with Equation (4.127) to give a relationship between incident, refracted and reflected wave forces. The corresponding relationship between displacements may be obtained by using Equations (4.130), i.e. from Equation (4.130a)

$$F = E_1 A_1 \frac{\partial u}{\partial x} = -\frac{E_1 A_1}{c_1} \frac{\partial u}{\partial t} \tag{4.132}$$

4.8.6.4 *Analysis of stress induced by piling*

In order to develop a simplified theory to obtain the stresses induced in a pile during the driving operation, an idealization of the hammer-pile system must be established. The method which is proposed reduces the problem to an equivalent 5DOF system and thus greatly simplifies the calculation of stress distribution within a pile. In view of the considerable difficulties in estimating the values for the various physical quantities involved in the phenomenon of piling, a simplified method of analysis, will be of considerable use to the engineer, provided it is used with care.

Smith (1962) has proposed a more complex model to represent the pile driving operation. The various components of the system are represented by weights and springs thus forming many degrees of freedom. A digital computer is used to obtain the response of the system. This method is more versatile than that being proposed herein, but involves more compution.

Figure 4.15 illustrates the general arrangement of the three main components of the piling operation. The hammer or ram is usually a short, rigid, heavy object and is represented by the weight $w = mg$. The dolly which is inserted between the hammer and the pile is usually a short, light springy object and is represented by a spring of stiffness without mass. The pile is a heavy object with elasticity through which longitudinal waves may propagate.

Three forces act upon the hammer during impact, the hammer weight, w, the inertia of the hammer, $m\ddot{x}$, and the reaction of the spring upon the hammer, F, which is also the pile-driving force. The equilibrium of the forces may be written as

$$F = m(g - \ddot{x}) \tag{4.133}$$

During impact the spring is compressed an amount x while the end of the pile is displaced by an amount u. Thus the following compatibility may be

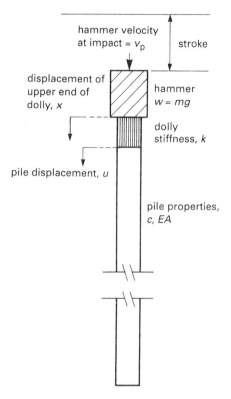

Fig. 4.15 General arrangement of the piling operation.

written

$$x - u = \frac{F}{k} \tag{4.134}$$

Rearranging, differentiating and substituting of Equation (4.134) into Equation (4.133) gives

$$F = m\left(g - \ddot{u} - \frac{\ddot{F}}{k}\right) \tag{4.135}$$

The driving force on the pile, which is opposite in sense to the reaction on the hammer, may be expressed in terms of the velocity of the pile $\partial u / \partial t$ as follows

$$F = -\sigma_x A = -EA\frac{\partial u}{\partial x} \tag{4.136}$$

in which σ_x is the stress in a pile of area A and modulus of elasticity E. Substituting Equation (4.130) into Equation (4.136) gives

$$F = \frac{EA}{c}\frac{\partial u}{\partial t} \tag{4.137}$$

which is combined with Equation (4.135) to give

$$\frac{EA}{kc} m\ddot{u} + m\ddot{u} + \frac{EA}{c} \dot{u} = mg \tag{4.138}$$

Equation (4.138) may be simplified by considering Equation (4.133). In practice the inertial part of Equation (4.133), i.e. the product $m\ddot{x}$, is very much greater than mg. If the term mg is, therefore, ignored in this equation then Equation (4.138) becomes

$$\ddot{v} + 2\omega\xi\dot{v} + \omega^2 v = 0 \tag{4.139}$$

where $\omega^2 = k/m$, $2\omega\xi = kc/EA$ and the velocity v has replaced $\dot{u}$.

Equation (4.139) is the familiar form of the SDOF equation with the velocity as the unknown. The solution to this equation (see Equation (3.18)) is

$$v = e^{-\xi\omega t}(C_1 \sin \omega_D t + C_2 \cos \omega_D t) \tag{4.140}$$

The constants C_1 and C_2 are evaluated from the initial conditions of pile velocity and accelerations. Since the pile is initially stationary then

$$v = \frac{\partial u}{\partial t} = 0 \quad \text{at } t = 0 \tag{4.141}$$

Substitution into Equation (4.140) gives $C_2 = 0$. If it is assumed that the velocity of the hammer at impact is v_p, then from Equation (4.138)

$$v_p = \dot{u} + \frac{\dot{F}}{k} = \frac{\dot{F}}{k} \tag{4.142}$$

since $\dot{u} = 0$ at $t = 0$. Combining Equations (4.142) and (4.137) and substituting $2\omega\xi = kc/(EA)$ gives the following equation for the initial pile acceleration

$$\dot{v} = 2\omega\xi\, v_p \tag{4.143}$$

Substituting into Equation (4.140) gives

$$A = 2\frac{\omega}{\omega_D}\xi v_p$$

and the equation may now be written as

$$v = e^{-\xi\omega t}\, 2\frac{\omega}{\omega_D}\xi v_p \sin \omega_D t \tag{4.144}$$

and the corresponding driving force from Equation (4.137) is

$$F = e^{-\xi\omega t}\frac{v_p k}{\omega_D}\sin \omega_D t \tag{4.145}$$

Chapter Five
Non-linear Dynamic Analysis

5.1 Introduction

The consideration of dynamic analysis which has been undertaken so far has assumed that the structures are linearly elastic. The principle of super-position will, therefore, hold. Such assumptions are not applicable in all cases of dynamic analysis. For example, the mathematical representations of damping and stiffness may be non-linear functions. A framed structure in which some members have yielded during the application of a dynamic load is an example of a structure in which stiffness changes have occurred as a result of the yielding and the associated analysis must be non-linear.

Non-linear analysis may be carried out with the aid of a computer. A convenient procedure is the step-by-step integration technique representing the non-linear behaviour of the system as a series of successively changing linear steps. In this method, the response of the system is determined for a series of time increments and dynamic equilibrium is established at the beginning and end of each increment. The non-linear properties of the system are modified at the beginning of each step and the values of velocity and displacement, which have been computed at the end of a particular increment, become the initial values at the start of the next step.

5.2 Incremental equilibrium equations

Consider the SDOF system which is shown in Fig. 5.1. During an interval of time Δt, the change in the applied force $\Delta p(t)$ will be balanced by corre-sponding changes in the inertia, damping and spring forces given by $m\,\Delta\ddot{x}$, $c(t)\,\Delta\dot{x}$ and $k(t)\,\Delta x$ respectively. The mass m is assumed to remain constant and $c(t)$ and $k(t)$ represent the damping and the stiffness for the time interval. The term Δx is the displacement increment corresponding to Δt and $\Delta\dot{x}$ and $\Delta\ddot{x}$ are the velocity and acceleration respectively. The incremental equilibrium equation may be written as

$$m\,\Delta\ddot{x} + c(t)\,\Delta\dot{x} + k(t)\,\Delta x = \Delta p(t) \tag{5.1}$$

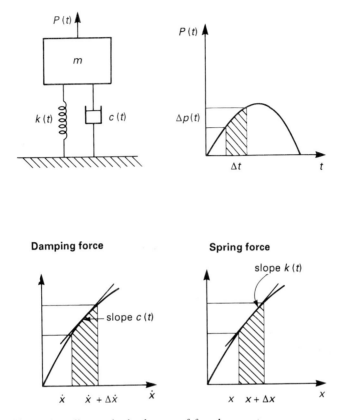

Fig. 5.1 An arbitrary non-linear single degree of freedom system.

Any kind of non-linearity may be represented in this equation, moreover, the mass need not remain constant. The values of $c(t)$ and $k(t)$ which correspond to a particular increment are taken to be the slopes of the damping force/velocity graph and spring force/displacement graph respectively for the increment (Fig. 5.1).

5.3 Integration of the incremental equilibrium equation

A step-by-step integration will be used to evaluate the non-linear Equation (5.1). It will be assumed that within each increment the acceleration varies linearly and that, therefore, the velocity varies parabolically and the displacement varies as a cubic (Fig. 5.2). It is further assumed that all other dynamic properties of the system remain constant within the increment $t + \Delta t$. The velocity and displacement increments can therefore be written as

$$\Delta \dot{x}(t) = \ddot{x}(t)\,\Delta t + \Delta \ddot{x}(t)\,\frac{\Delta t}{2} \tag{5.2}$$

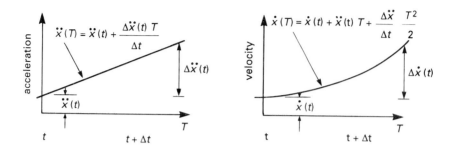

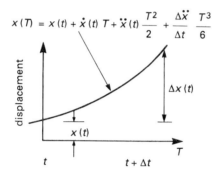

Fig. 5.2 The assumed variations of acceleration, velocity and displacement for a step-by-step integration.

$$\Delta x(t) = \dot{x}(t) \, \Delta t + \ddot{x}(t) \, \frac{\Delta t^2}{2} + \Delta\ddot{x}(t) \, \frac{\Delta t^2}{6} \tag{5.3}$$

From Equation (5.3)

$$\Delta\ddot{x}(t) = \frac{6}{\Delta t^2} \, \Delta x(t) - \frac{6}{\Delta t} \, \dot{x}(t) - 3\ddot{x}(t) \tag{5.4}$$

and

$$\Delta\dot{x}(t) = \frac{3}{\Delta t} \, \Delta x(t) - 3\dot{x}(t) - \frac{\Delta t}{2} \, \ddot{x}(t) \tag{5.5}$$

Substitution of Equations (5.4) and (5.5) into Equation (5.1) yields the following equation of motion in which the displacement increment is the variable

$$m\left[\frac{6}{\Delta t^2} \, \Delta x(t) - \frac{6}{\Delta t} \, \dot{x}(t) - 3\ddot{x}(t) \right] + c(t)\left[\frac{3}{\Delta t} \, \Delta x(t) - 3\dot{x}(t) - \frac{\Delta t}{2} \, \ddot{x}(t) \right]$$
$$+ k(t) \, \Delta x(t) = \Delta p(t) \tag{5.6}$$

Equation (5.6) may be rearranged in the form of a dynamic stiffness relationship given by

$$\Delta \bar{p}(t) = \bar{k}(t)\,\Delta x(t) \tag{5.7}$$

in which

$$\bar{k}(t) = k(t) + \frac{6}{\Delta t^2}\,m + \frac{3}{\Delta t}\,c(t) \tag{5.8}$$

$$\Delta \bar{p}(t) = \Delta p(t) + m\left[\frac{6}{\Delta t}\,\dot{x}(t) + 3\ddot{x}(t)\right] + c(t)\left[3\dot{x}(t) + \frac{\Delta t}{2}\,\ddot{x}(t)\right] \tag{5.9}$$

Equation (5.9) contains the known initial conditions. Equation (5.7) is solved for the incremental displacement and the substitution of this into Equation (5.5) enables the incremental velocity to be obtained. These incremental values are added to the values of the velocity and displacement at the beginning of the increment to form the initial values for the next increment.

Errors could accumulate from increment to increment as a result of the linear assumption for the variation of acceleration and the further assumption of constant dynamic properties within a time increment. This accumulation of errors is avoided by using the total equilibrium relationship at the beginning of each time step, thus equating the total external load to the total damping and elastic forces. The acceleration may then be expressed as

$$\ddot{x}(t) = \frac{1}{m}\left[p(t) - f_D(t) - f_s(t)\right] \tag{5.10}$$

where $p(t)$ is the external load and $f_D(t)$ and $f_s(t)$ are the damping and elastic forces respectively.

5.4   Computer program for the analysis of a non-linear system

The step-by-step integration technique for the dynamic analysis of a SDOF non-linear system may be conveniently programmed to run on a small personal computer. What follows is a description of such a program in which a listing, flow charts and a sample analysis have been provided.

The program is capable of analyzing the response of a SDOF system with non-linear stiffness and either viscous or hysteretic damping. The dynamic loading versus time curve may take any form. In particular, the results of actual dynamic measurements may be used since the program presents the dynamic loading as a least squares fit to a Fourier sine series representation.

The engineer may use the program directly by simply reproducing the listing in a form suitable as input to the computer or he/she may wish to

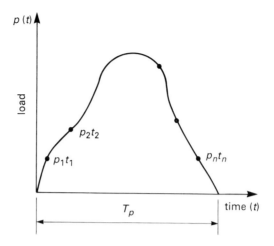

Fig. 5.3 An arbitrary pulse of dynamic loading versus time.

implement modifications such as the inclusion of a particular non-linear damping relationship. In this latter case the engineer must study the listing and flow charts in detail. Such modifications are likely to be straightforward.

Before discussing the program, an appropriate outline will be given of the Fourier series representation and the least squares fit to be used.

5.4.1 Fourier sine series representation of dynamic loading

Figure 5.3 shows a single impulse of dynamic load, $p(t)$, versus time, t. The figure also shows n sets of load−time data given as (p_1, t_1), (p_2, t_2), ..., (p_n, t_n). The single impulse may be represented by a Fourier sine series of the form

$$p(t) = C_1 \sin at + C_2 \sin 2at + \ldots + C_m \sin mat \qquad (5.11)$$

where $m \leq n$ and $a = \pi/T_p$, T_p being the duration of the pulse.

5.4.2 Least squares curve fit

If the n sets of data shown in Fig. 5.3 are substituted into Equation (5.11)

$$\sum_{i=1}^{m} c_i f_i(t) - p_j = \delta_j \qquad j = 1, 2, \ldots, n \qquad (5.12)$$

where the δ_js are the residuals, $f_i(t) = \sin(iat)$, and it is required to choose the c_is such that the sum of the squares of the residuals is a minimum, i.e.

$$\sum_{j=1}^{m} \delta_j^2 = \text{minimum value} \qquad (5.13)$$

Since Equation (5.13) is a function of the c_is, a necessary minimum condition is that the partial derivatives with respect to each of the c_is must equal zero, i.e.

$$\frac{\partial}{\partial c_k} \sum_{j=1}^{m} \delta_j^2 = 0 \qquad \text{for } k = 1, 2, \ldots, m \qquad (5.14)$$

or

$$\sum_{j=1}^{m} \delta_j \frac{\partial \delta_j}{\partial c_k} = 0 \qquad (5.15)$$

From Equation (5.12)

$$\frac{\partial \delta_j}{\partial c_k} = f_k(t_j) \qquad (5.16)$$

Substituting Equations (5.12) and (5.16) into Equation (5.15) and rearranging gives

$$\sum_{i=1}^{m} c_i \sum_{j=1}^{n} f_i(t_j) f_k(t_j) = \sum_{j=1}^{n} p_j f_k(t_j) \qquad (5.17)$$

Equation (5.17) is a set of n linear equations for the m unknowns $c_1, c_2, \ldots, c_m$ and they may be solved by a Gaussian reduction

$$\sum_{i=1}^{m} a_{ki} c_i = b_k \qquad (5.18)$$

where $\qquad a_{ki} = \sum_{j=1}^{n} f_i(t_j) f_k(t_j) \qquad (5.19)$

and $\qquad b_k = \sum_{j=1}^{n} p_j f_k(t_j) \qquad (5.20)$

The following matrix notation is used to express Equations (5.18), (5.19) and (5.20)

$$\mathbf{F} = \begin{bmatrix} f_1(t_1) \; f_1(t_2) \ldots \ldots \ldots f_1(t_n) \\ \vdots \\ f_m(t_1) \ldots \ldots \ldots \ldots f_m(t_n) \end{bmatrix}$$

$$\mathbf{B} = \begin{bmatrix} b_1 \\ \vdots \\ b_m \end{bmatrix} \qquad \mathbf{P} = \begin{bmatrix} p_1 \\ \vdots \\ p_m \end{bmatrix} \qquad \mathbf{C} = \begin{bmatrix} c_1 \\ \vdots \\ c_m \end{bmatrix}$$

$$A = \begin{bmatrix} a_{11} \ a_{12} \ldots \ldots \ldots a_{1m} \\ \vdots \\ a_{m1} \ldots \ldots \ldots \ldots a_{mm} \end{bmatrix}$$

hence, Equation (5.18) may be written as

$$A \ C = B \tag{5.21}$$

where $A = F \ F^{T}$ $\tag{5.22}$

and $B = F \ P$ $\tag{5.23}$

Substituting Equations (5.22) and (5.23) into Equations (5.21) gives

$$F^{T} \ C = P \tag{5.24}$$

The least squares solution to the above system of equations, where there may be more equations than unknowns (n equations, m unknown c_is), is found by premultiplying both sides of Equation (5.14) by F and solving the resulting set of m equations and m unknowns by a Gaussian reduction.

5.4.3 The computer program

The following procedure is implemented within the computer program for any given time increment.

(1) Initial values of velocity $\dot{x}(t)$ and displacement $x(t)$ are known either as initial conditions or from the values at the end of the previous increments.
(2) The damping $c(t)$ and stiffness $k(t)$ can be obtained for a particular increment from the non-linear properties of the structure. These values combine with $\dot{x}(t)$ and $x(t)$ to give the current values of $f_D(t)$ and $f_s(t)$.
(3) The initial acceleration $\ddot{x}(t)$ is obtained by Equation (5.10).
(4) The effective load increment $\Delta p(t)$ and the effective stiffness Δk are given by Equations (5.8) and (5.9).
(5) The displacement increment $\Delta x(t)$ is obtained from Equation (5.7) and the velocity increment $\Delta \dot{x}(t)$ from Equation (5.5).
(6) The velocity and displacement at the end of each increment are obtained from

$$\dot{x}(t + \Delta t) = \dot{x}(t) + \Delta \dot{x}(t) \tag{5.25}$$

$$x(t + \Delta t) = x(t) + \Delta x(t) \tag{5.26}$$

The above procedure may be repeated for any number of time increments to give the response history of the system. A linear problem may be solved if constant values of damping and stiffness are used in step (2).

The accuracy of the method depends upon the length of the time increment Δt. To obtain a convergent solution it has been found that acceptable results may be obtained if the ratio $\Delta t / T$ is less than 0.1 where T is the natural period of vibration of the structure.

The flow chart for the complete program is shown in Fig. 5.4 and the following list is a key to the program variables.

5.4.3.1 Variable names in program for SDOF structure

Program MOVE

Real numbers

PH (15)	=	load $P_1, \ldots, P_n$ ⎫ loading data
X(15)	=	times $t_1, \ldots, t_n$ ⎭
R	=	b, gives the periodicity for Fourier sine series
B(10)	=	coefficients $C_1, \ldots, C_m$ in Fourier sine series
SK	=	stiffness constant in kN/m
SC	=	damping coefficient in kN/m
SM	=	mass in kg
FSY	=	yield force in kN
TK	=	stiffness in kN/m
FS	=	elastic force
FD	=	damping force
P	=	load at time t
PP	=	load at time $t + \delta t$
VV(350)	=	displacements stored for drawing the graph
V	=	displacement in m
VW	=	displacement in mm
DV	=	velocity in m/s
DVW	=	velocity in mm/s
DDV	=	acceleration in m/s^2
CV	=	change in displacement over one time increment
CDV	=	change in velocity over one time increment
T	=	time
DT	=	time increment, Δt
TR	=	total time of analysis of response
OK	=	$\bar{K}$, the equivalent static stiffness
SCR	=	⎫ factors for hysteretic damping
SKSC	=	⎭

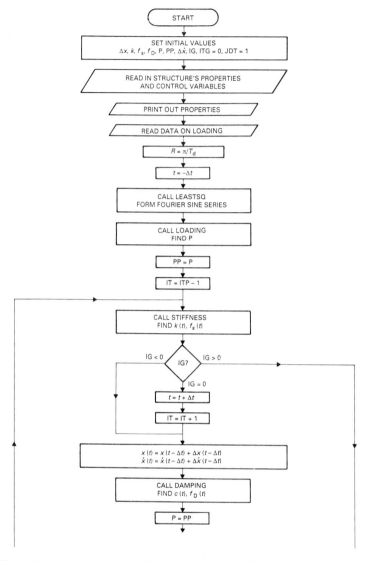

Fig. 5.4   Flow chart for the analysis of a non-linear SDOF system.

DF		$= \ \mathrm{PP} - \mathrm{P} = \Delta p$
ODP		$= \ \Delta\bar{p}$, the change in equivalent static loading over Δt

Integer numbers

N		$=$ number of sets of load data, n
ITP		$=$ number of time increments to be skipped on print-out

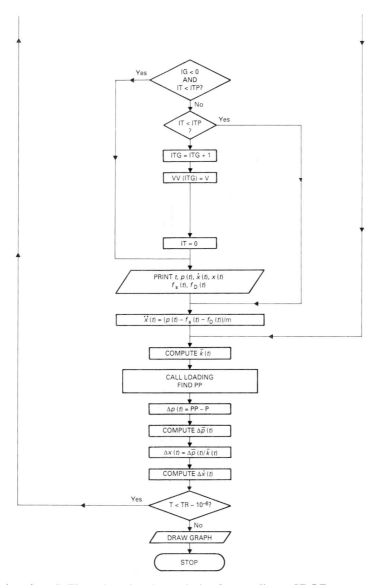

Fig. 5.4 (continued) Flow chart for the analysis of a non-linear SDOF system.

IT	=	count for ITP
NFS	=	number of terms in Fourier sine series, m
ITG	=	controls input to VV(ITG), (initially $= 0$)
IG	=	controls return from STIFFNESS when sub-increments are used, (initially $= 0$)
JDT	=	controls forming of subincrements, (initially $= 1$)

LINE(120)	=	array used to form each line of graph	
BLANK, DOT, CROSS	=	alphanumerics used to plot graph	graph
K	=	cross position on graph	

Subroutine GAUSS

M	= number of unknowns
A(10,10)	= known coefficients
B(10)	= input data, which when solved, become the unknown coefficients $c_1, c_2, \ldots, c_m$

Subroutine LEASTQ

Y(15)	= PH(15)
X(15)	= X(15)
F(10,15)	= **F**
A(10,10)	= **A**
B(10)	= **B**
G	= GAUSS

Subroutine STIFFNESS

FSA	= trial value of f_s
DDT	= δt_N, when DT is δt_s
TT	= trial value for t in subincrement

Subroutine LOADING

TL	= T
B(10)	= $c_1, c_2, \ldots, c_m$

The following are the important stages in the organization of the computing procedure.

(1) The program reads the input data and prints the system properties. The period of a single pulse, R_1, is then calculated.

(2) The subroutine LEASTQ is called to form the matrices **A** and **B** of section 5.4.2. LEASTQ then calls the subroutine GAUSS which solves Equation (5.24) by a Gaussian reduction to give the coefficients in the matrix **C** which are stored in the matrix **B**.

(3) The subroutine LOADING is next called to calculate the initial load which is normally zero.

(4) The cycle for each increment of time begins by calling the subroutine STIFFNESS which calculates the elastic properties and force. If this subroutine is modelling an elastoplastic system with an abrupt yield point, it is necessary to subdivide the time increment into two parts, each with constant properties corresponding to the pre-yield and post-yield conditions. To achieve this, an iterative procedure is

required to establish the length of each subincrement. This procedure is included in STIFFNESS and is described in the next section. The subroutine STIFFNESS is placed at the start of the cycle in order to establish if the structure yielded in the last standard time increment.

(5)   The stiffness and elastic forces are obtained and the time, velocity and displacement are updated by adding the current increments to the previous values.

(6)   The subroutine DAMPING is called which calculates the damping coefficient and damping force for the current increment. The initial loading for this increment is then set equal to the final loading of the previous increment.

(7)   A number of statements are now passed which decide if the output from the current increment should be printed or skipped. This facility is provided since it is unlikely that the user will require output for every time increment.

(8)   The acceleration is calculated using Equation (5.10) and the effective stiffness using Equation (5.8).

(9)   The subroutine LOADING is called to calculate the load at the end of the time increment and then the effective load increment is calculated using Equation (5.9).

(10) The displacement and velocity increments are calculated from Equations (5.7) and (5.5) respectively. This completes the incremental calculations and a check is made to ensure that the analysis has reached its final point in time.

5.4.3.2   Stiffness, damping and loading subroutines

The STIFFNESS and DAMPING subroutines may be modified to suit particular elastic and damping properties.

The STIFFNESS subroutine
This routine accommodates elastoplastic behaviour and uses the bisector method to subdivide the increment within which the structure yields. The flow diagram for the routine is shown in Fig. 5.5, and Fig. 5.6 shows ideal elastoplastic behaviour.

Using Fortran notation, the elastic force is given by

$$FSA = FS + TK * CV$$

where FSA is the updated elastic force, FS is the elastic force from the previous increment, TK = SK is the linear elastic stiffness and CV is the change in displacement during the previous time increment.

If FSA > FSY or < −FSY, then the new value of FS = ±FSY, whichever is appropriate, and the stiffness TK = 0. If the structure is still elastic,

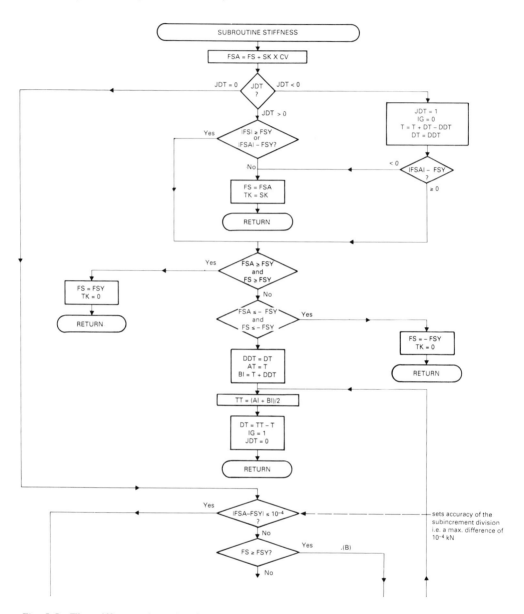

Fig. 5.5 The stiffness subroutine flow chart for elastoplastic behaviour with subincrements. Initially IG = ITG = 0, JDT = 1.

−FSY < FSA < FSY and the new values of FS = FSA and TK = SK, the elastic stiffness.

After FSA has been calculated, JDT is tested. JDT normally equals 1, but may equal 0 or −1 during the calculation of subincrements. If it is assumed that JDT = 1, then FS and FSA are tested to establish if they are

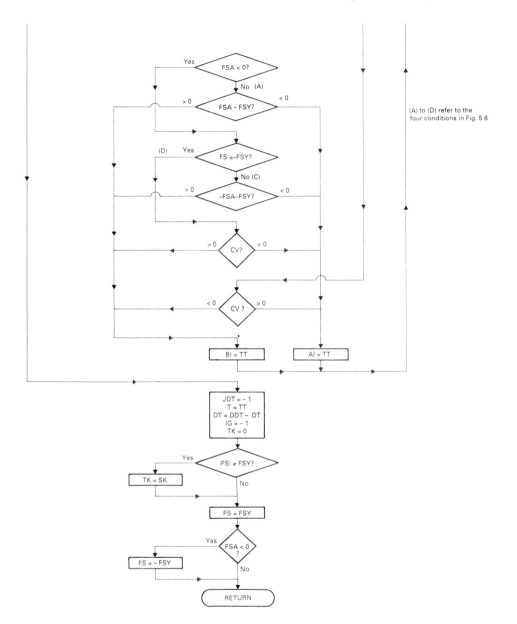

Fig. 5.5 (continued) The stiffness subroutine flow chart for elastoplastic behaviour with subincrements. Initially IG = ITG = 0, JDT = 1.

> FSY or < −FSY. Assuming that the situation is linear so that FS = FSA and TK = SK, control then passes to the master segment. If this last condition is satisfied, then plastic behaviour is occurring. The next test determines if both FS and FSA are ≥ FSY or ≤ −FSY, in which case the new FS = ±FSY accordingly and TK = 0.

If neither of the above conditions are satisfied, then one of the conditions identified as A, B, C, D or E must occur, where yield starts or finishes. The corresponding time increment is subdivided using the bisector method and this technique is described with the aid of Fig. 5.7 which is applicable to A and E of Fig. 5.6.

At $t = T + DT$, FSA $>$ FSY and it is required to find FSA $=$ FSY. Now another value of t is tried

$$t = \frac{T + (T + DT)}{2} = T + \frac{DT}{2}$$

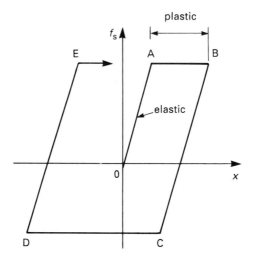

Fig. 5.6   Elastoplastic behaviour.

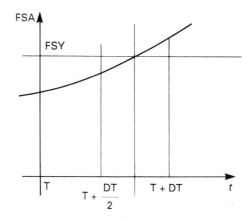

Fig. 5.7   Principle of the bisector method.

in this case FSA < FSY. Hence a large t is required. The next attempt would be

$$t = \left[\left(T + \frac{DT}{2}\right) + (T + DT)\right]\frac{1}{2} = T + \frac{3DT}{4}$$

FSA − FSY is very small and is assumed to be within an acceptable tolerance. Therefore, the new time increment is given by DT = 3DDT/4 where DDT is the standard increment which has been defined at the start of the program. The process may be repeated until sufficient accuracy has been achieved.

A similar procedure is applicable to the conditions B, C and D of Fig. 5.6 where the curve of FSA moves in different directions relative to FSY or −FSY.

Returning to the section in the STIFFNESS subroutine which implements the bisector method, DDT = DT, hence retaining the standard time increment. AI and BI are the earliest and latest times respectively to be subdivided. Initially, AI = T and BI = T + DDT. The bisected time, TT = (AI + BI)/2 is calculated and the new trial interval becomes DT = TT − T.

At this stage the stiffness, damping, time, forces, acceleration, velocity and displacement are unchanged as the last increment is now to be re-run with a different DT, and FSA will be compared with FSY. To enable this to be carried out IG = 1, control is passed to the master segment and the statements which update the incremented variables mentioned above are skipped.

JDT = 0 so that |FSA − FSY| is tested on return to the STIFFNESS subroutine. If a further iteration is required with a different value of DT, the procedure follows a series of IF statements to determine which conditions from A to D (E is the same as A) of Fig. 5.6 are appropriate and then continues, to determine whether DT has been over- or underestimated. Then either AI or BI = IT and the process is repeated.

When the condition |FSA − FSY| is sufficiently small, JDT = −1, TT = T and DT = DDT ensuring that the sum of the two subincrements equals the standard increment. IG = −1, so that the time T is not updated on return to the master segment, since this has been done in the subroutine. Two IF statements are then used to select the appropriate values of TK and FS, the new value of FS = ±FSY, and control passes to the master segment so that the second subincrement can be considered. On returning to STIFFNESS, JDT = −1 which requires JDT being reset to 1, IG to 0, T = T + DT − DDT and DT = DDT which returns the time increment to its standard value.

The procedure is now ready to return to the standard time increments, but since FS = ±FSY, one further IF statement is required to direct the procedure back to an appropriate statement ensuring that the correct values for TK and FS are used.

The DAMPING subroutine

The program listing provides two alternative DAMPING subroutines depending on whether the damping is considered to be viscous or hysteretic. Only the appropriate lines must be retained, the others must be removed before the program is run.

For viscous damping, the damping force is given by the Fortran statement

$$FD = FD + CVD* SC$$

For hysteretic damping, the magnitude of the damping force is calculated from

$$f_D = \xi k \frac{f_s}{k} \frac{\dot{x}}{|\dot{x}|} \tag{5.27}$$

A fuller description of the mechanisms of structural damping has been given in section 3.9 of Chapter 3.

The LOADING subroutine

This subroutine determines the dynamic load, P, by summing the sine functions multiplied by their coefficients, $B(I)$, which have been calculated in the subroutine LEASTQ. An IF statement checks when the time reaches the maximum impulse time and sets $P = 0$ from this time onwards.

In its current form, the program can only deal with a single impulse. Should the dynamic loading consist of a number of cycles of pulses, the analysis may be conducted in stages with the response values at the end of one pulse becoming the initial values for the next pulse. This means that the program is re-run for each pulse with the appropriate change in the input data.

5.4.3.3 The program listing

A complete Fortran listing of the program now follows. If the problem involves viscous damping, the statements associated with hysteretic damping (these are indicated) should be removed. Similarly, if the problem has hysteretic damping, the subroutine DAMPING related to the viscous case should be removed.

```
C
C   DYNAMIC ANALYSIS OF A NON-LINEAR SINGLE DEGREE OF
C   FREEDOM STRUCTURE
C
C
      CHARACTER BLANK,DOT,CROSS,LINE
      DIMENSION PH(15),X(15),B(10),VV(350),LINE(120)
      DATA CV,TK,FS,FD,P,PP,CDV/7*0./,IG,ITG,ODT/0,0,1/
```

```
      OPEN(1,FILE='DATA')
      OPEN(2,FILE='RESULT')
  101 FORMAT('0','SYSTEM PROPERTIES')
  102 FORMAT('0','MASS(KG)',13X,'=',F8.2)
  103 FORMAT('0','TIME INTERVAL(S)',5X,'=',F8.6)
  104 FORMAT('0','INITIAL DISP.(M)',5X,'=',F8.6)
  105 FORMAT('0','INITIAL VEL.(M/S)',4X,'=',F8.6)
  106 FORMAT('0','DAMPING COEF(KNS/M)=',F8.5)
  107 FORMAT('0','STIFFNESS(KN/M)',6X,'=',F8.3)
  108 FORMAT('0','YIELD FORCE(KN)',6X,'=',F8.3)
  109 FORMAT('0','TIME(S)',2X,'LOAD(KN)',2X,'DISP(MM)',2X,
     'VEL(MM/S)1',2X,'FS(KN)',4X,'FD(KN)')
  110 FORMAT('  ',F6,3,F10,4,F10,3,F11,3,F11,4,F11,4)
  111 FORMAT(3I0)
  112 FORMAT('  ',3A1)
  113 FORMAT('0','2',19X,'1',19X,'0',19X,'11',19X,'2',19X,'3')
  114 FORMAT('  ',120A1)
      READ(1,*)SM,SC,SK,FSY,V,DV,TR,DT
      READ(1,*)N,ITP,NFS
      WRITE(2,101)
      WRITE(2,102)SM
      WRITE(2,103)DT
      WRITE(2,104)V
      WRITE(2,105)DV
      WRITE(2,106)SC
      WRITE(2,107)SK
      WRITE(2,108)FSY
      READ(1,*)(PH(I),I=1,N)
      READ(1,*)(X(I),I=1,N)
C
C  R GIVES THE FUNCTIONS OF THE FOURIER SINE SERIES THE
C  DESIRED
C  PERIODICITY TO REPRESENT THE APPLIED LOADING
      R=3.141593/(X(N)=X(1))
C
C     COEFFICIENTS(B=MATRIX)FOR THE FOURIER SERIES ARE
      FOUND BY THE
C     METHOD OF LEAST SQUARES AND GAUSSIAN REDUCTION
      CALL LEASTSQ(PH, X,N,B,R, GAUSS, NFS)
      WRITE(2,109)
      SCR=3.141593*SC*SQRT(SK*1000/SM)/2
      SKSC=SCR/SK
      T=-DT
      XN=X(N)
      CALL LOADING(P,B,R,T+DT,XN,NFS)
      PP=P
```

```
        IT=ITP−1
C
C       START OF LOOP TO CALCULATE NEW SITUATION FOR EACH
C       TIME INCREMENT, DT
    1   CALL STIFFNESS(FSY,FS,CV,SK,TK,T,DT,JDT,IG)
        IF(IG)6,7,9
    7   T=T+DT
        IT=IT+1
    6   V=V+CV
        DV=DV+CDV
        CALL DAMPING(SKSC,FS,DV,FD)
        P=PP
        IF(IG.LT.0.AND.IT.LT.ITP)GOTO 8
        IF(IT.LT.ITP)GOTO 2
        ITG=ITG+1
        VV(ITG)=V
        IT=0
    8   VW=V*1000
        DVW=DV*1000
        WRITE(2,110)T,P,VW,DVW,FS,FD
    2   DDV=(P−FD−FS)*1000/SM
    9   OK=TK+(6*SM/(DT*1000)+3*SC)/DT
        CALL LOADING(PP,B,R,T+DT,XN,NFS)
        DP=PP−P
        ODP=DP+SM*(6*DV/DT+3*DDV)/1000+SC*(3*DV−DT*DDV/2)
        CV=ODP/OK
        CDV=3*CV/DT−3*DV−DT*DDV/2
        IF(T.LT.(TR−1E−6)GOTO 1
C
C       READ PLOTTING SYMBOLS
        READ(1,*)DOT,CROSS,BLANK
C
C       PRINT A LINE OF DOTS, WHICH WILL BE VERTICAL AXIS WHEN
C       PAGE IS TURNED
        WRITE(2,113)
        DO 3 J=1,120
    3   LINE(J)=DOT
        WRITE(2,114)LINE
C
C       PRODUCE GRAPH
        DO 4 J=1,120
    4   LINE(J)=BLANK
        DO 5 I=1, ITG
        LINE(41)=DOT
        K=250*VV(I)+41.5
        LINE(K)=CROSS
```

```
      WRITE(2,114)LINE
  5   LINE(K)=BLANK
      STOP
      END

      SUBROUTINE LEASTSQ(Y,X,N,B,R,G,NFS)
      DIMENSION Y(15),X(15),F(10,15),A(10,10),B(10)
      DO 20 J=1,N
      DO 20 I=1,NFS
 20   F(I,J)=SIN(I*R*X(J))
      DO 22 I=1,NFS
      DO 22 K=1,I
      A(K,I)=0
      DO 21 J=1,N
 21   A(K,I)=A(K,I)*F(I,J)*F(K,J)
 22   A(I,K)=A(K,I)
      DO 23 K=1,NFS
      B(K)=0
      DO 23 J=1,N
 23   B(K)=B(K)+Y(J)*F(K,J)
      CALL GAUSS(A,NFS,B)
      RETURN
      END

      SUBROUTINE GAUSS(A,M,B)
      DIMENSION A(10,10),B(10)
      DO 30 I=1,(M−1)
      DO 30 J=(I+1),M
      FAC=A(J,I)/A(I,I)
      DO 31 K=I,M
 31   A(J,K)=A(J,K)−A(I,K)*FAC
 30   B(J)=B(J)−B(I)*FAC
      B(M)=B(M)/A(M,M)
      DO 32 I=1,(M−1)
      J=M−I
      DO 33 L=(J+1),M
 33   B(J)=B(J)−A(J,L)*B(L)
 32   B(J)=B(J)/A(J,J)
      RETURN
      END

      SUBROUTINE LOADING (P,B,R,TL,XN,NFS)
      DIMENSION B(10)
      P=0
      IF(TL.GE.XN)GOTO 41
      DO 40 I=1,NFS
```

```
40 P=P+B(I)*SIN(I*R*TL)
41 RETURN
   END

   SUBROUTINE DAMPING (FD,CDV,SC)
   FD=FD+CDV*SC
   RETURN
   END

   SUBROUTINE DAMPING (SKSC,FS,DV,FD)
   IF(DV)60,61,60
61 FD=0
   GOTO 63
60 FD=ABS(FS*SKSC)*DV/ABS(DV)
63 RETURN
   END

   SUBROUTINE STIFFNESS(FSY,FS,CV,SK,TK,T,DT,JDT,IG)
   FSA=FS+SK*CV
   IF(JDT)15,17,19
19 IF(ABS(FS).GE.FSY.OR.ABS(FSA).GE.FSY)GOTO   18
53 FS=FSA
   TK=SK
   RETURN
18 IF(FSA.GE.FSY.AND.FS.LE.(-FSY))GOTO 12
   IF(FSA.LE.(-FSY).AND.FS.LE.(-FSY))GOTO 13
   DDT=DT
   AI=T
   BI=T+DDT
10 TT=(AI±BI)/2
   DT=TT-T
   IG=1
   JDT=0
   RETURN
12 FS=FSY
   TK=0
   RETURN
13 FS=-FSY
   TK=0
   RETURN
17 IF(ABS(FSA-FSY).LE.1E-4)GOTO 11
   IF(FS.GE.FSY)GOTO 52
   IF(FSA.LT.0)GOTO 50
   IF(FSA-FSY)14,14,16
50 IF(FS.LE.(-FSY))GOTO 51
   IF(-FSA-FSY)14,14,16
```

```
51  IF(CV)14,14,16
52  IF(CV)16,16,14
14  AI=TT
    GOTO 10
16  BI=TT
    GOTO 10
11  JDT=−1
    T=TT
    DT=DDT−DT
    IG=−1
    TK=0
    IF(ABS(FS).GE.FSY)TK=SK
    FS=FSY
    IF(FSA.LT.0)FS=−FSY
    RETURN
15  JDT=1
    IG=0
    T=T+DT−DDT
    DT=DDT
    IF(ABS(FSA)−FSY)53,18,18
    END
```

5.5 Example

The following problem should enable the reader who wishes to implement the computer program to test the results. Figure E5.1 shows a portal frame made from steel universal columns supporting a rigid horizontal floor thus allowing joint rotations to be considered negligible. The modulus of elasticity, E, may be taken as $206\,kN/mm^2$.

It is assumed that the beam supports a dead loading of $4\,kN/m^2$ from an area which is 5 m square. The equivalent mass is, therefore, $10^4\,kg$, and the column masses are neglected. The horizontal stiffness of the portal frame is $24EI/h^3$, and for a 203×203 universal column, the stiffness $k = 5\,907\,kN/m$.

The yield force p_y causing the formation of plastic hinges at the column ends is assumed to be $200\,kN$.

For undamped free vibrations, $\omega = \sqrt{(k/m)} = 24.304$ radians/second and the natural period, $T = 2\pi/\omega = 0.259\,s$. Therefore the critical damping $c_c = 2m\omega = 486\,kN/m$. For steel beams, the damping ratio, γ, approximately equals 0.10 and the damping factor $c = \gamma c_c = 4.9\,kN/m$.

Figure E5.1 shows the impulse of dynamic loading $p(t)$ which is to be applied to the structure. In order to exceed the yield stress, a maximum load of $225\,kN$ has been chosen. The impulse lasts for one second and twelve load−time data points have been specified.

If p_y is taken as $500\,kN$, then an elastic response of the structure will be

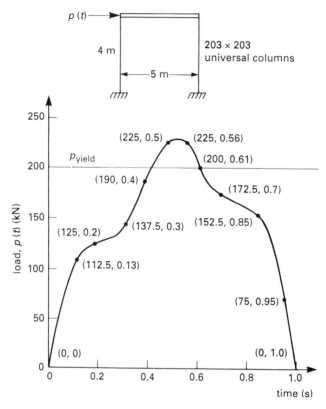

Fig. E5.1   Example problem of a single-storey portal frame subjected to an impulsive horizontal load $p(t)$.

obtained. The time increment Δt has been taken as 0.01 s. A discussion of hysteretic damping has been given in section 3.9 of Chapter 3. In order to ensure approximately the same energy absorption per cycle for both viscous and hysteretic damping, the following calculation has been adopted for this example problem.

The hysteretic damping force is given by

$$f_\mathrm{D} = \xi k \, |x| \, \frac{\dot{x}}{|\dot{x}|}$$

where the hysteretic damping coefficient

$$\xi = \frac{\pi c}{2\sqrt{(mk)}}$$

and

$$\xi k = \frac{c\pi}{2} \sqrt{\left(\frac{k}{m}\right)}$$

The product ξk is considered to be as a constant and is calculated at the beginning of the program, i.e. $\xi k = 187\,\text{kN/m}$. The damping force is then calculated according to Equation (5.27) from which it can be seen to be directly proportional to the elastic force. Thus excessive damping is avoided if yielding has taken place and the equation

$$f_\text{D} = \xi k \,|x|\, \frac{\dot{x}}{|\dot{x}|}$$

is used.

5.5.1 Data input

Six lines of data input are required. The lines are written in free format according to the following instructions:

Line 1 (Structure properties)

mass	damping coefficient	stiffness	yield force
SM (kg)	SC (kN/m)	SK (kN/m)	FSY (kN)

Line 2 (Initial conditions and limits)

initial disp.	initial vel.	time range	time interval
V (m)	DV (m/s)	TR (s)	DT (s)

Line 3

N	ITP	NFS
no. of sets of load−time data	no. of answers to be skipped on print-out	no. of terms to Fourier series

Note: NFS $\leq$ N

Line 4
PH(I)   list of loads (kN)

Line 5
X(I)   list of times (s) corresponding to PH(I)

Line 6
(dot, cross, blank)   graphical plot facility

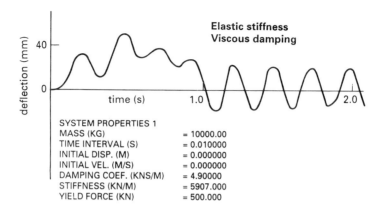

SYSTEM PROPERTIES 1
MASS (KG)	= 10000.00
TIME INTERVAL (S)	= 0.010000
INITIAL DISP. (M)	= 0.000000
INITIAL VEL. (M/S)	= 0.000000
DAMPING COEF. (KNS/M)	= 4.90000
STIFFNESS (KN/M)	= 5907.000
YIELD FORCE (KN)	= 500.000

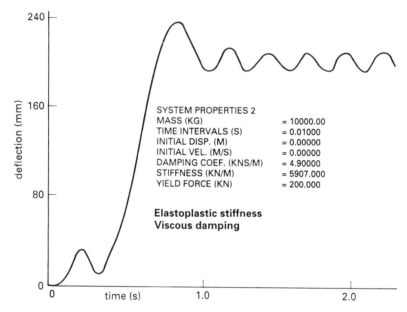

SYSTEM PROPERTIES 2
MASS (KG)	= 10000.00
TIME INTERVALS (S)	= 0.01000
INITIAL DISP. (M)	= 0.00000
INITIAL VEL. (M/S)	= 0.00000
DAMPING COEF. (KNS/M)	= 4.90000
STIFFNESS (KN/M)	= 5907.000
YIELD FORCE (KN)	= 200.000

Fig. E5.2 The response of the viscously damped system of Fig. E5.1 with elastic and elastoplastic stiffness.

The actual data for the six lines used for this example are as follows:

```
IE4  4.9  5907  200
2  0  10  0.001
12  5  8
0  112.5  125  137.5  190  225  225  200  172.5  152.5  750
0  0  13  0.2  0.3  0.4  0.5  0.56  0.61  0.7  0.85  0.95  1.0
.X
```

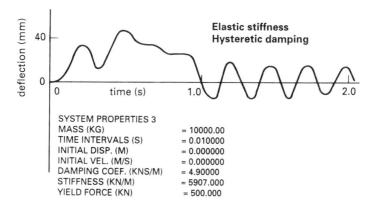

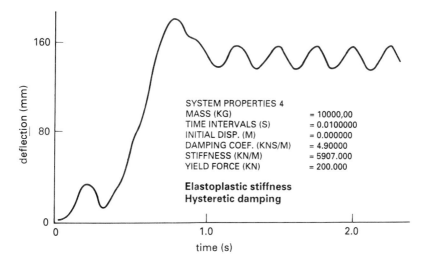

Fig. E5.3 The response of the hysteretically damped system of Fig. E5.1 with elastic and elastoplastic stiffnes.

5.5.2 Results

Figures E5.2 and E5.3 show the first two seconds of the response of the system shown in Fig. E5.1. Figure E5.2 compares the elastic and elastoplastic response of the system with viscous damping. The effect of plastic yielding can be seen in both cases by the formation of a permanent set about which subsequent free vibrations take place.

Although the results represent the response of an arbitrary SDOF system, they demonstrate the way in which the program may be applied.

In order to convince himself that the foregoing principles have been fully understood, the reader should implement the program involved.

Chapter Six
Vibration Instrumentation and Data Analysis Techniques

6.1 Introduction

Vibration problems frequently occur in practical engineering situations. Even though allowances may be made in design for dynamic behaviour, many more problems are actually experienced in practice than are predicted or explained by theory. It is for this reason that the role of vibration measurement and the associated analysis of data are particularly important in obtaining the dynamic characteristics of structural response.

There are basically two types of vibration measurement, each requiring the accurate determination of vibration displacement or a function of displacement. These are

(1) The measurement of vibration levels or of the dynamic response of a structure or structural component in its normal service environment, and
(2) The measurement of the absolute vibration properties of a structure or structural component.

Measurements of the first kind involve recording, for immediate or subsequent analysis, the vibration levels which a structure experiences when subjected to a dynamic excitation.

In a civil or structural engineering context, this type of measurement involves a full-size structure such as a tall building, a bridge or off-shore structure, for example, subjected to a relevant dynamic source such as traffic, machinery, shock or ocean waves. The reasons for the measurements might be to check the real dynamic behaviour of a structure in which the analysis and design have provided for the effects of vibration. Alternatively, the structure may already be functional before a source of vibration appears. A piling operation adjacent to an existing structure would be such an example and in this case nuisance and damage thresholds require assessment.

Measurements of this kind obviously cannot be observed under laboratory conditions and it is extremely unlikely that the true nature and magnitude of the exciting forces would be known exactly. Only the total or overall response of the structure to the excitation source is obtained. This is, of course, very important since it is often the only information available. However, it should be appreciated that in this case only a limited amount of information is obtained regarding the dynamic characteristics of the structure in these conditions.

The second kind of measurement is more comprehensive. The structure is excited in a controlled manner by a source which is simple and defined. During the test the input and the response to input are both measured and it is possible to determine the required dynamic properties of the structure. Such tests are generally preferred.

Before measurements are taken, it is important to possess some knowledge of the vibrations being measured and the form in which the resulting data is required. This information will assist in determining the methods of measurement and data reduction as well as in the selection of the associated instrumentation.

In this context it is helpful to identify, albeit somewhat generally, the types of forced vibration to which structures may be subjected. There are broadly three classes

(1) Harmonic vibration, such as machinery vibrations, consisting of frequencies of which the higher values are multiples of the lower frequencies.
(2) Shock-induced vibrations, such as the structural response induced by piling or blasting.
(3) Random vibration which may be caused by earthquakes or by the motion of ocean waves, for example.

In practice, many vibrations are a combination of either two or even three of the classes. A wide range of equipment is available for the analysis of these vibrations.

6.2 Relevant experimental response quantities

In order to define vibratory motion, either totally or in part, it is necessary to obtain the appropriate experimental measurements. In principle, measurements may be taken to obtain

(1) Natural frequencies and mode shapes
(2) Specific dynamic properties, such as stiffness and damping
(3) Magnitudes of acceleration, velocity and amplitude

(4)   Total number of cycles for fatigue analysis.

Since it has been stated that it may not be possible to obtain all dynamic data, it is necessary to select the appropriate instrumentation to observe the relevant response quantities for a given situation.

Structural engineers are particularly interested in strain and the measurement of dynamic strain is usually obtained without difficulty. It is also possible to obtain a measurement of the relative displacement between two points. It will be seen that the principles of vibration-measuring equipment can be explained in terms of strain and relative displacement measurements.

6.3   Measuring instruments

At a time when developments in electronics are influencing measurement technology, the engineer is often confronted with a wealth of technical literature. Although these developments improve the electronic details of dynamic measuring instruments, basic principles do not change. In describing measuring instruments, the author has adopted an approach in which basic principles and instrument theory are explained and only where necessary has consideration been given to electronic details.

Due attention will be given to the important practical details concerning the instruments to be described and the various types of transducer which might be used. It is intended that the information will enable the engineer to evaluate the suitability of a particular measuring device.

Vibration measuring instruments consist of two fundamental parts; the mechanical part, which follows the motion being measured and the transducer which converts relative motion into an output form. In some instruments such as crystal accelerometers and strain gauge devices, the two parts are inseparable.

The majority of vibration-measuring instruments are absolute measurement instruments such as the mass-spring devices which are often referred to as seismic instruments. These consist of a mass-spring-damper system attached to a base which vibrates with the structure being investigated. A transducer measures the relative displacement between the base and the mass. Crystal accelerometers are also seismic instruments but the spring, which is the crystal, is also the transducer. The absolute measurements are independent of a fixed reference point.

The alternative to the absolute measuring instrument is the fixed reference device in which motion of the vibrating structure is measured relative to some fixed point. Instruments whose mode of operation is based upon optical or moving-coil principles are included in this category. In fact, the operating principles of such instruments are often the same as those of the transducers in seismic devices.

6.4 The principles of seismic instruments

The fundamental principles of seismic instruments can be demonstrated by reference to Fig. 6.1, which represents a typical schematic arrangement. The mass is connected through the parallel spring and damper arrangement to the base of the instrument which is in contact with the vibrating source.

Since the mass tends to remain fixed in space, the motion is registered as a relative displacement between the mass and the base. This displacement is sensed and registered by an appropriate transducer.

The seismic instrument may be used for either displacement or acceleration measurement, the particular mode of operation depending upon the magnitude of the mass and the spring stiffness. Generally, a large mass and a soft spring are desirable for the measurement of displacement, whereas a relatively small mass and a stiff spring are used for the measurement of acceleration.

If the base of the mass-spring-damper system is subjected to a motion u, and the relative motion of the base and mass which is sensed by the transducer is x, then the differential equation of motion is given by

$$m\ddot{x} + c\dot{x} + kx = -m\ddot{x} \tag{6.1}$$

where m, c and k are the mass, damping constant and spring stiffness respectively.

If it is assumed that $u = u_0 \cos \omega t$ and if the transient part of the solution to Equation (6.1) is neglected, then $x = x_0 \cos(\omega t - \theta)$ where u_0 and x_0 are the maximum values of u and x respectively, θ is the phase angle, ω the circular frequency and t is the time. As a result

$$\frac{x_0}{u_0} = \left(\frac{\omega}{\omega_n}\right)^2 \bigg/ \sqrt{\left\{\left[1 - \left(\frac{\omega}{\omega_n}\right)^2\right]^2 + \left[2\xi\left(\frac{\omega}{\omega_n}\right)\right]^2\right\}} \tag{6.2}$$

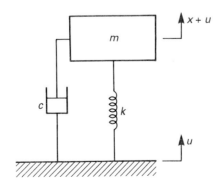

Fig. 6.1 Schematic mass–spring–damper system.

$$\theta = \tan^{-1}\left[\frac{2\xi(\omega/\omega_n)}{1 - (\omega/\omega_n)^2}\right]$$ (6.3)

where ω_n is the undamped natural circular frequency of the system given by

$$\omega_n = \sqrt{(k/m)}$$ (6.4)

and ξ is the ratio of the damping coefficient to the critical damping coefficient given by

$$\xi = c/\sqrt{(2km)}$$ (6.5)

Figure 6.2 shows the variation of the ratio of amplitude x_0 to the amplitude u_0, plotted against (ω/ω_n) for a range of damping ratios. In Fig. 6.3 the variation of the normalized ratio of amplitude x_0, to the base acceleration amplitude $\ddot{u}_0$, which is equal to $-\omega^2 u_0$, is plotted against (ω/ω_n).

The ratio x_0/u_0 is almost a constant at frequencies above the natural frequency and the ratio $\omega_n^2 x_0/\omega^2 u_0$ is almost a constant below the natural frequency for a damping ratio approximating to 0.6. Thus if it is assumed that the transducer produces a measurement of relative displacement, then the displacement of a structure is measured for frequencies above the natural frequency.

For measurements taken at frequencies below the natural frequency of a transducer, the relative displacement is proportional to the acceleration of a structure and for this latter case the instrument is called an accelerometer. Usually displacement-measuring seismic instruments have low natural frequencies (approximately 1 hertz) and accelerometers have high natural frequencies (greater than 50 hertz and often as high as 100 kilohertz).

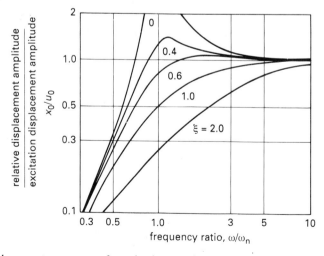

Fig. 6.2 Displacement response of a seismic transducer.

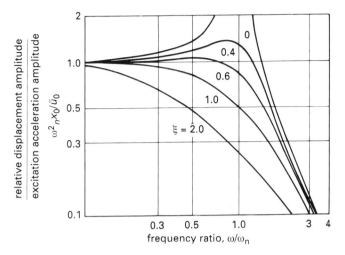

Fig. 6.3   Acceleration response of a seismic transducer.

Therefore, displacement-measuring instruments are bulky and heavy and acceleration-measuring instruments are relatively light.

Some instruments are capable of being fitted with springs of different stiffness and can measure both displacement and acceleration. The transducer can take a variety of forms; it can measure the velocity x_0, if, for example, it consists of a moving coil within a magnet. In this case, if the instrument is a displacement device, then the output will be proportional to the velocity of the structure. Alternatively, an instrument with an accelerometer movement produces an output which will be proportional to the rate of change of acceleration.

Further modifications in the role of these instruments may be obtained by incorporating integrating and differentiating amplifiers to operate on the output signals enabling a variety of functions to be obtained from a single instrument.

6.5   Practical details of seismic instruments

Notwithstanding the influence of rapid developments in electronics, reliable results can be obtained using simple mechanical devices. This is particularly true in the case of civil and structural engineering environments, where it might be desirable to use robust instrumentation rather than sophisticated electronic devices.

However, whatever instrumentation is finally selected, certain details of terminology are common and it is useful to consider these as a practical basis for information.

Instrument sensitivity is defined as the ratio of available output to the mechanical input quantity. In the case of a displacement-measuring instrument, the sensitivity would be expressed in millivolts per metre.

Instruments are usually designed to measure vibration in one direction only. However, some response will be observed when an instrument is excited in a direction which is perpendicular to the usual direction of measurement. This effect is termed cross-axis sensitivity and should be as small as possible. Cross-axis sensitivity is caused by the non-alignment of the movement between the indicated direction of measurement and the axes which are perpendicular to this direction. For crystal accelerometers the cross-axis sensitivity depends upon the crystal properties.

The resolution of an instrument is important, since it defines the smallest change in measured quantity which can be detected. In mechanical devices the resolution is governed by the friction of the moving parts. However, this limit is much less than the limit imposed by the transducer and associated electrical parts. These latter limits would include resolution of meter reading, electrical noise and the ability of the transducer to sense very low levels of resistance or capacitance, for example.

There are clear limits for the range of operation of a seismic instrument and these are defined by an operating envelope, which is a graph of displacement, velocity and acceleration versus frequency, see Fig. 6.4. Limit lines are drawn on this graph defining an operating envelope within which the instrument is functional.

The maximum displacement limit is usually determined by mechanical stops, while the maximum acceleration is determined by the severity of the inertia loading upon the instrument. Instrument resolution usually governs the lower limits of displacement and acceleration. It has been stated in

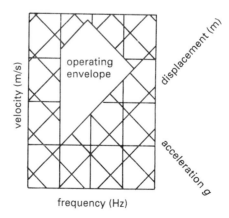

Fig. 6.4 An example of an operating envelope for an instrument.

section 6.4 that frequency limits govern the mode of operation of seismic instruments. For displacement instruments the upper limit is usually governed by secondary resonances, while for accelerometers the lower limit of frequency is usually determined by the characteristics of the transducer or of the associated amplifiers.

Consideration should be given to the phase shift which is introduced by the dynamics of the movement. For the working frequency ranges of displacement and acceleration, the phase angle is non-linearly related to the frequency except when the damping ratios are zero and 0.65 approximately. Disregarding the case of zero damping, displacement transducers can never reproduce exactly the time records of complex displacements, because the components of the complex frequency signal are delayed by different amounts. However, for accelerometers with a damping ratio of 0.65 a near-linear relationship exists between phase angle and frequency so that the component time delay is identical for measurement purposes. Hence an accelerometer will produce accurate records of acceleration for complex signals.

Some further factors which affect response are temperature, humidity, acoustic noise and magnetic fields. Manufacturers usually provide details concerning these effects to facilitate quantification.

6.6   Transducers

Electrical methods of detecting vibration are most widely utilized and the conversion of a mechanical motion into an electrical signal may be achieved in different ways. This section considers the modes of operation of the common types of transducer which convert mechanical vibration into an electrical signal.

6.6.1   Variable resistance transducer

Variable resistance transducers require an externally supplied voltage to energize them and they may be classified as slider or strain gauge types.

6.6.1.1   Slider type

In this transducer, the slider moves across a resistor and taps a voltage which is proportional to the change of resistance and hence to the displacement. The resistance may be continuous such as a single wire, or stepped such as a coil and can be measured by several techniques, the best known being the Wheatstone bridge principle. This device is formed of four identical arms one of which is the transducer resistor. The bridge is balanced when the amplitude is zero and no potential difference exists between the arms of the bridge. Under conditions of vibration, the motion of the slider is converted into a potential difference which is proportional to displacement. The slider type is not usually incorporated in a seismic device.

6.6.1.2 Strain gauge types

Strain gauges consist of a conducting element, which is fixed to the vibrating surface of an object to measure the strain associated with the motion. The resistance of the gauge changes when it is subjected to strain. There are several types of gauge, the most common being the foil gauge which is fully bonded to a surface.

Important factors affecting performance are the gauge factor and the resistance and power-handling capacities of the gauge. The gauge factor is the ratio of the fractional change of resistance to the fractional change of length. Values approximately equal to 2.0 are usual. Since the gauge requires an external voltage for operation, power is dissipated and the amount depends upon the gauge resistance. The voltage that can be applied to a gauge depends upon the power dissipated at normal working temperatures. The power-handling capability of a gauge should be sufficiently low to avoid unstable behaviour. Manufacturers' details will normally consider these points.

The Wheatstone bridge circuit is used for the measurement of strain and has particular advantages in the case of dynamic strains, when pairs of gauges can be arranged in opposite arms of the bridge. For example, gauges may be bonded to either side of a vibrating cantilever to measure the alternating strain very accurately.

Strain gauges may be used as the transducers in seismic instruments, particularly for accelerometers, when they may operate at zero frequency. However, a disadvantage is their very low output voltage requiring amplification without interference.

Piezo-resistive strain gauges are crystals whose resistance changes with strain. A particular advantage of this type of strain gauge is the very high gauge factor (up to 150) and hence its sensitivity, enabling very small strains to be measured.

6.6.2 Variable capacitance transducer

The principle of the variable capacitance transducer is based upon the detection of mechanical motion by the change of capacitance between capacitor plates. This type of transducer is a displacement-measuring device and is capable of detecting smaller vibrations than the variable resistance type. It has excellent amplitude and frequency ranges with high resolution. The associated electronics, however, are more complex than for the variable resistance transducer.

The capacitance should vary almost linearly with displacement in well-designed transducers. The presence of non-linear effects is due to the capacitance being inversely proportional to the distance between capacitor plates and also to capacitive plate edge effects.

Capacitance can be detected using several different associated circuits. However, a common technique is to use an a.c. circuit incorporating a carrier wave, whose frequency is several times that of the frequency of the measured displacement. Two alternatives are possible whereby the change of capacitance modifies the carrier either by amplitude modulation or frequency modulation. The capacitance-measuring circuit is different for each case.

6.6.3 Variable inductance transducer

There are a number of transducers which employ the property of inductance to detect displacement. Of these, the differential transformer transducer is perhaps the most widely used for vibration work.

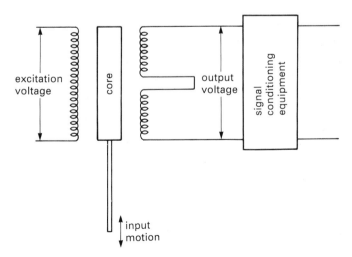

Fig. 6.5 Simplified representation of the differential transformer.

In this type, a cylindrical high permeability core is positioned within an assembly of three coaxial cores. There is no mechanical contact between these components. Normally, the primary coil is equidistant from the two secondary coils. Voltages are induced in the two secondary coils when an a.c. or d.c. voltage is used to energize the primary coil. Usually an a.c. voltage is preferred since it provides a higher order of accuracy, linearity and stability but at a greater cost to the overall system. The secondary coils are connected differentially. If the higher permeability core is moved from the null position, the voltage increases in one secondary core and decreases in the other, resulting in an output voltage which is due to the subtraction of voltages in opposition. Figure 6.5 shows a general circuit for such a transducer. The distance and direction of the movement are determined by the electrical output.

The transducer is capable of detecting extremely small relative displacements between the high permeability core and the three core assembly. It is particularly useful for low frequency work since it is capable of operating down to zero frequency.

6.6.4   Electro-dynamic transducer

The principal of the electro-dynamic transducer is different from that of the transducers considered so far. Instead of producing an electrical signal proportional to displacement, the signal produced by an electro-dynamic transducer is proportional to velocity. In this velocity transducer, a conductor moves in a fixed magnetic field and a voltage is generated which is proportional to the rate of cutting of the field. Alternatively, the conductor may remain stationary while the magnet producing the magnetic field is allowed to vibrate. The conductor is usually a circular coil wound on a light former which can move within an annular magnetic field. The form of the transducer can be modified to measure rotational vibrations. Figure 6.6 shows a simplified diagram of an electro-dynamic transducer.

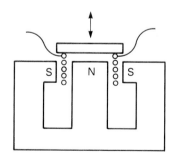

Fig. 6.6 Simplified representation of an electro-dynamic transducer.

The transducer self-generates the power needed and requires no power supply. Damping may be provided by generating eddy currents in the conducting former or by placing a low-valued resistance across the output from the circular coil around the conductor. The linearity and low frequency response of electro-dynamic transducers are acceptable although not as good as other types of transformer.

6.6.5   Piezo-electric or crystal transducer

The piezo-electric effect is the phenomenon associated with the generation of an electrical charge across two faces of a crystal, when the crystal is subjected to strain. It is a further property of certain types of crystal.

The strain applied to an appropriate crystal causes the internal electrical equilibrium to be disturbed and it is only regained when electrical charges of opposite sign collect on opposite faces of the crystal.

A variety of crystals exhibit the piezo-electric effect, some occurring naturally while others are synthetic. Further types are manufactured from polarized ceramic materials. Of the natural crystals, quartz rochelle salt and tourmaline are typical. Ammonium dihydrogen phosphate is a common synthetic crystal while barium titanate, lead metaniobate and lead zirconate titanate are typical polarized ceramic compounds. The crystals have different qualities which may make some types more suitable in certain environments. For instance, quartz crystals may be used for high temperature applications whereas the ceramic compound type, being more sensitive, may be used to detect exceedingly small amplitudes of motion. Ceramic crystals have a better low-frequency response and are cheaper to produce than other crystal types. However, they depolarize with time and, therefore, their response characteristics change accordingly.

In seismic instruments, the piezo-electric crystals act as both the transducer and spring in the movement. Since the crystal is very stiff, the natural frequency is high and, therefore, the instrument responds as an accelerometer.

Piezo-electric accelerometers are commonly used and Fig. 6.7 shows a typical cross-section through such a transducer. A single crystal or a series of crystals may be cut in a direction which is most sensitive to the direction of operation of the transducer to produce the most effective electrical charge.

6.6.6   The transistor transducer

Transistors are available which may be used to sense pressure or point

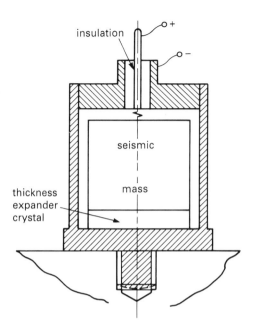

Fig. 6.7 Cross-section through a piezo-electric accelerometer.

forces. The Pitran PT2 and PT3 transistor types are examples. These transducers are silicon n−p−n planar versions with the emitter−base junction mechanically coupled to a diaphragm at the top of the sealing cylinder. If a pressure or point force is applied to the diaphragm a reversible change in the characteristics of the transducer is produced. The device may be used in acceleration-measuring instruments.

6.7 Indicating and recording instruments

It is the usual practice to record the electrical signals produced by vibration-measuring instruments and to analyze these records later. Certain information can be obtained at the instant of measurement, such as the indicated level of maximum displacement. The sophisticated techniques of data analysis, which are to be discussed, require expensive equipment and are often time-consuming. Unless the measurement of vibration is associated with a laboratory, it is unlikely that conditions would be suitable for analysis in most civil and structural engineering situations at the time of measurement. The necessity to record the effects of vibration for subsequent analysis is thus important. Those instruments in common use will be considered in what follows.

6.7.1 Oscillographs

An oscillograph is basically an instrument for recording the movements of a galvanometer. This movement is related to the amplified output of a transducer. For the recording of vibration effects, two types of galvanometer movement are available. These are the direct-writing stylus movement and the light-beam movement. Typical arrangements of these movements are illustrated in Fig. 6.8.

The stylus movement records the galvanometer deflection directly by tracing on recording paper which is passing the stylus at a pre-determined rate. A time variation of the movement is thus produced. The recording may be made by ink writing, pressure writing or heated styli. The latter two use pressure or heat-sensitive paper.

The light-beam movement which takes the place of the stylus movement incorporates the principle of the light beam galvanometer in which a fine beam of light is produced to follow the galvanometer movement and to record on light-sensitive paper.

The frequency response of an oscillograph is determined by the response of the associated galvanometer and this depends upon the mass moment of inertia. Since the mass of the stylus movement is greater than that of the light beam movement, the frequency response of the latter is much higher. Generally, stylus type movements exhibit frequency responses from 0 to 100 hertz while light-beam oscillographs respond up to 50 kilohertz. Movements are damped to about 0.6 of critical in order to obtain the widest band

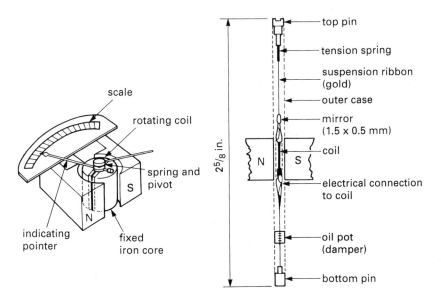

Fig. 6.8 Typical arrangements of indicating and light beam galvanometer movements.

of flat response. Oscillographs may provide single or multiple channels of information.

6.7.2 Cathode-ray oscilloscope

In this instrument an electron gun provides a beam of electrons, which move in a vacuum towards a fluorescent screen where the beam is displayed as a light spot. The beam passes between a pair of horizontal plates and a pair of vertical plates. If the signal from a transducer is connected to the vertical plates, the associated electrostatic field deflects the electron beam by an amount proportional to the transducer signal. If the horizontal input plates are then connected to a sweep circuit, the beam will move at a constant horizontal velocity across the screen and trace the time variation of the transducer signal.

It is usual to attach a camera to the screen and record the motion of the light spot. Time records are conveniently obtained by passing photographic paper continuously across the screen, in which case the sweep circuit is not required and the light spot moves in the vertical direction only. Transient motions can easily be recorded using a triggering facility. A still or Polaroid camera is left with its shutter open; the signal of the transient motion triggers the light spot on its path across the screen and the camera records the motion.

6.7.3 Magnetic tape recorders

Magnetic tape recorders possess many advantages that cannot be reproduced by other types of recorder. Certain advantages are attributable to the magnetic medium, such as being able to play back and reproduce the recording innumerable times with minimum deterioration. Erasure and recording on the same medium are other advantages of magnetic recording. The extended frequency range, linearity and accuracy of tape recorders are a result of engineering development.

The basic mode of operation is that a recording head magnetizes the tape in proportion to the input signal and a pick-up head senses the associated change of magnetic flux recorded on the tape. The type of recorder suitable for civil and structural engineering purposes is likely to be expensive because of the need to extend the recorded frequency response down to zero frequency. Frequency modulation techniques are used to achieve this important requirement. The upper frequency limit of magnetic tape recorders may be as high as 100 kilohertz with a flat response and this is achieved using amplitude modulation techniques.

Finally, the output signal from a tape recorder can be conveniently used as the input to other recording or analysing instruments. Recorders have a multichannel record and replay facility.

6.8   Vibration and shock-generating equipment

In order to evaluate the dynamic response of a structure or structural component, it may be desirable to conduct dynamic tests in which the source of vibration is produced by electrical or mechanical equipment. These tests may be conducted in the field, such as the dynamic excitation of a multi-storey building, or they may be conducted in the laboratory on an appropriate model. In either case, the equivalent source of vibration should represent, to an acceptable degree of accuracy, the actual source which is being reproduced. Many types of device exist which are capable of producing harmonic, random and shock sources of vibration.

6.8.1   Harmonic generators

A sinusoidal motion of constant maximum displacement may be produced mechanically by a device similar to that shown in Fig. 6.9.

Alternatively, the crank mechanism may be replaced by a cam or a yoke. Such arrangements are ideal for producing constant maximum displacement vibrations at low frequencies.

The reaction type vibration generator is an alternative mechanical form. The type which has been used most effectively for the dynamic testing of buildings consists of two unbalanced rotating masses, as in Fig. 6.10. Each mass rotates in a direction opposite to the other, such that unbalanced forces, which are proportional to the square of the frequency of rotation, are additive in the desired direction of vibration. Since the masses may be

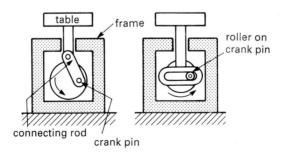

Fig. 6.9 Mechanism for producing harmonic vibrations.

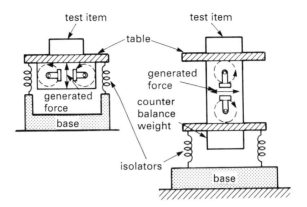

Fig. 6.10   Mechanism for producing harmonic vibration.

varied, the device may be used over operating ranges for both force and frequency.

Electro-dynamic generators are capable of producing constant forces up to relatively high frequencies. They are bulky, however, and possess a lower frequency limit below which they cannot operate effectively.

Electro-dynamic generators are also capable of producing constant forces which may be very large. These devices possess a significant upper frequency limit and a minimum displacement below which the effects of friction are important. These latter generators are generally used for impedance (see section 6.13).

6.8.2   Random generators

Electro-dynamic devices may be used as random vibration generators. In fact, such generators are capable of producing most practical types of vibration wave form if the input to the device is suitably conditioned through filters and amplifiers. Thus, the vibrations from a random source, or indeed from any other source, may be recorded on magnetic tape and after suitable conditioning may be used as the input to an electro-dynamic generator which will reproduce the dynamic characteristics of the original source.

6.8.3   Shock-generating machines

It is extremely difficult and often impossible to simulate the shock motions which are likely to occur in practical situations. This is not a great disadvantage in civil or structural engineering, since in most cases dynamic performance may be adequately defined in harmonic or random test environments.

However, testing equipment does exist which may produce a variety of shock motions. These motions are produced by dropped weights, ballistic pendulums and mechanical hammers operating in a controlled manner to produce the desired shock characteristics.

6.9   An introduction to data analysis techniques and the associated instrumentation

6.9.1   Introduction

The objective of data analysis is to extract from the record of vibration the variation of some relevant parameter such as the amplitude of acceleration with both time and frequency. Reduction of data to the time domain is called magnitude-time analysis while reduction to the frequency domain is called spectral analysis. The reduced form of data is used either to predict the response of a structure to vibration from a source, thus requiring the source vibration to be measured, or to analyze the response of a structure if the measurements have been obtained from the structure itself. In the latter case it would also be desirable to analyze the source of vibration without the influence of the structure. It has already been pointed out that this may not be possible.

From an analysis of data it should be possible to describe a vibration in terms of its magnitude, frequency content and phase. This description may be simple, such as for harmonic or periodic motions, or it may be relatively complex, such as for the statistical descriptions associated with random vibrations.

Whatever the type of vibration, it is essential to know how a structure will respond. This response is given by the mode shapes and associated frequencies as well as by the magnitude of vibration. Other properties of the motion, such as its damping characteristics, should also be known. Appropriate methods of data analysis should be used to obtain the foregoing information.

An introduction to data analysis techniques and comments on the associated equipment are therefore presented. A discussion of vibration and impedance testing follows.

6.10   Harmonic and periodic vibration analysis

If a source of vibration excitation is harmonic or periodic, data analysis may be carried out using wave, spectrum or Fourier techniques. Such techniques would be appropriate to investigate the response of foundations supporting machinery or generators. Additionally, the harmonic testing of structures, which is an important method of determining resonant frequencies, would also use these techniques.

It has been stated that vibration excitation may be described by the variation of a parameter with both frequency and time. In the case of harmonic or periodic vibrations, the wave analyzer, which is an analogue device, obtains the variation of amplitude, for instance, of a particular frequency in the time-domain (Fig. 6.11).

This is obtained by selecting the desired frequency on the analyzer, which filters out all other frequencies from the input signal. The output from the analyzer represents the variation of amplitude with time at the desired frequency.

In practice, the electrical characteristics of the wave analyzer affect the result because it is not possible to isolate a single frequency for filtering and the resultant output will contain a small contribution from adjacent frequencies.

The objective of spectral analysis is to obtain the variation of a parameter with frequency. An electrical signal representing the vibration may be used as the input to a spectrum analyzer and the frequency selection filter swept electronically across the input. The output will be the variation of amplitude, for example, against a simultaneous display of the component frequencies (Fig. 6.11). The spectrum or frequency analyzer is also an analogue device.

The Fourier analyzer is a digital device which is much faster and more versatile than the wave and spectrum analyzers. It can be used to reduce vibration signals to both the time and frequency domains. A periodic

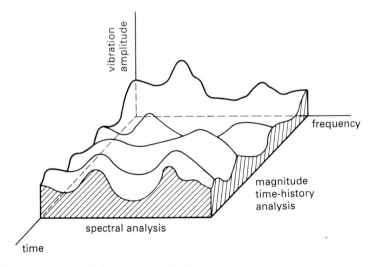

Fig. 6.11   Representation of vibration amplitude in the time and frequency domains.

function may be represented by a Fourier series and the coefficents in the series are a measure of the frequency content of the periodic function.

The Fourier analyzer utilizes this concept and converts the electrical input signal into digital form. The analyzer carries out the necessary operations to produce a number of functions, which are essential to vibration analysis, for example, parametric variation in time and frequency. The device may be used to analyze any of the three broad classes of vibration which have been previously mentioned in section 6.1.

Since most practical structures are complicated, a number of modes are likely to participate in the motion of the structure at a particular frequency. A direct reduction of data to the frequency domain can be misleading if the natural frequencies and the damping of the modes are required. This is due to the false zero caused by the motion and interference of other modes.

The Nyquist or polar is an alternative form of data reduction which may be used when resonant response is not controlled by a single frequency (Fig. 6.12).

The plot is obtained from a measurement of the total response in all modes at a point in a structure. This response is split into an in-phase component which is at a phase of 90° to the exciting force. The structure is assumed to possess hysteretic damping, which is a more realistic representation of internal structural damping. However, the plot may also be used with acceptable accuracy for systems containing small amounts of viscous damping.

An explanation of the Nyquist plot may be given if a single degree of

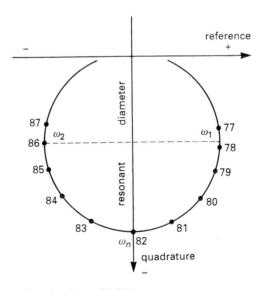

Fig. 6.12 Nyquist or polar plot for a SDOF system.

freedom system such as that shown in Fig. 6.1 is considered, for which the base motion $u = 0$ and with hysteretic damping replacing the viscous damping. Although this type of damping may be represented by a damping constant, which is inversely proportional to frequency, it is more conveniently described by a complex stiffness term given by $k(1 + i\eta)$, where η is the damping factor. In this case the equation of motion of the system subjected to a forced harmonic motion given by $pe^{i\omega t}$ may be written as

$$\ddot{x} + k(1 + i\eta)x = pe^{i\omega t} \tag{6.6}$$

then

$$x_0 = \frac{p}{k} \frac{\{[1 - (\omega/\omega_n)^2] - i\eta\}}{[1 - (\omega/\omega_n)^2] + \eta^2} \tag{6.7}$$

The static deflection is given by p/k and ω_n^2 equals k/m.

The real and imaginary components of x_0, relative to the force p, may be plotted on a complex diagram to form the Nyquist or polar plot. For the single degree of freedom system with hysteretic damping, it can be shown that the plot is a circle which is centred on the imaginary or phase-equals-90° axis and passes through the origin. The natural frequency of the system is known to be the lowest point on the circle and it is, therefore, unnecessary to take a measurement at this point. The diameter of the circle, which passes through the resonant frequency, is called the resonant diameter and the damping ratio of the mode of vibration is determined from the diameter which is perpendicular to the resonant diameter. The frequencies which correspond to this diameter are called the half power points ω_1 and ω_2. If the resonant frequency of the system is ω_n, it can be shown that

$$\eta = \frac{\omega_2 - \omega_1}{\omega_n} \tag{6.8}$$

For multi-degree of freedom systems, the Nyquist plot consists of a number of circles each offset from the origin by the amount of motion in the other modes. Figure 6.13 shows a Nyquist plot in which several frequencies are seen to overlap. It should still be possible to obtain the frequencies of interest for each circle or mode, and to use Equation (6.8). In cases when the polar plot contains a considerable amount of modal overlap the resulting diagram may be difficult to interpret. If this situation occurs it is advisable to repeat the experimentation using different positions of excitation and measurement to obtain improved results.

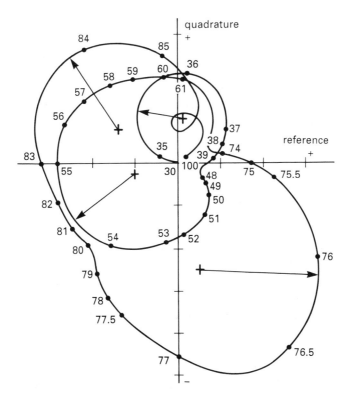

Fig. 6.13 Nyquist or polar plot for a MDOF system.

6.11 Shock data analysis

Shock, impulsive or transient excitations may last for a few milli-seconds or several seconds. Shock-induced vibrations have been described as one of the three broad classes of forced vibration and Chapter 2 considers some aspects of the dynamic response of structures subjected to impulsive or shock motions.

Blasting, piling operations and forging hammer impulses are common sources of shock. These vibrations are essentially complex motions which are likely to include a low or zero frequency content. For this reason, accelerometers are used for the measurement of shock motion. The output from an accelerometer, which is a time-history trace, can be conveniently recorded by a frequency-modulated recorder. The shock motion is thus preserved in a form suitable for subsequent data analysis. Generally, such a measurement, which represents an original time-history of motion or loading, is unsuitable for engineering purposes and some reduction to a more suitable form is required. The form of reduced data depends upon the ultimate use to which it is to be put and consideration will now be given to methods of shock data reduction.

6.11.1   The reduction of shock data

In order to provide shock data in a form which is particularly suited to engineering applications, two basically different approaches may be followed in describing the shock.

(1)   The properties of the shock may be described in the frequency or time domain
(2)   The properties of the shock may be described in terms of the effect of the shock upon a structure. This latter representation is referred to as a reduction to the response domain.

6.11.2   The reduction of data to the frequency domain

A non-periodic complex shock may be represented by the superposition of sinusoidal components, each component having its own magnitude and phase. In order to describe a shock motion in terms of its sinusoidal components and their phases, a Fourier spectrum analysis is conducted. It is not intended to discuss the theory of this type of analysis, but only to mention that the Fourier spectrum is a plot of the magnitude or phase of the sinusoidal components against frequency and is computed numerically for complex shock motions. The Fourier spectrum is analogous to the Fourier components of a periodic function. Figure 6.14 shows the form that a Fourier spectrum might take. Discrete values may be taken from the spectrum and used to excite a model structure or an equivalent electrical analogue of a structure.

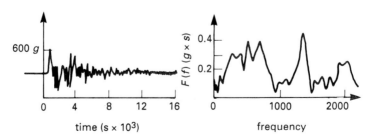

Fig. 6.14   A Fourier spectrum plot of a shock motion.

6.11.3   Reduction of data to the response domain

A dynamic system will have a characteristic response to a shock input excitation. Certain effects of the shock upon the system may be defined by relating the response peaks to the properties of the system. By limiting the dynamic system to a linear, viscously-damped, single degree of freedom structure with lumped parameters, the response depends upon the undamped natural frequency and the fraction of critical damping.

The shock or response spectrum is used to demonstrate how a simple idealization of a structure will respond to a shock excitation. The two-dimensional spectrum is a plot of the maximum peak response versus the natural frequency of the system. The two-dimensional spectrum may be extended to three dimensions by including the number of peak excursions exceeding a particular level. The response of a system may be defined in terms of any one of the parameters which describe its motion. The more usual forms of spectra, however, use displacement or acceleration. Figure 6.15 illustrates typical shock spectra. It should be noted that the original time-history of the shock cannot be reproduced from a knowledge of the peak responses of the shock spectra.

Shock spectra are usually obtained by employing electrical analogue or numerical methods, the procedure being very lengthy.

An electrical analogue of the equations of motion of the single degree of freedom structure can be represented on an analogue computer. A frequency-modulated tape recorded would provide the analogue computer with a time-history record of the shock motion. The computer will output the response motion from which the peak values may be obtained. Changes in natural frequency and damping can be achieved conveniently.

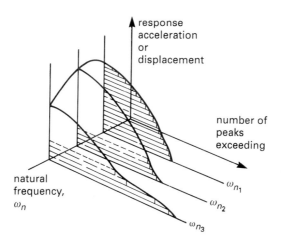

Fig. 6.15   A typical three-dimensional shock spectrum.

Numerical methods of obtaining shock spectra are based upon the step-by-step evaluation of Duhamel's integral using a digital computer, and some examples have been given in Chapter 3. The computer can count the peaks in a response and produce a graphical output of the analysis.

A simple practical technique which may be used to obtain a shock spectrum is to use a reed gauge. This instrument consists of small flexible cantilevers supporting different end masses, hence each 'reed' has a different natural frequency. Styli at the end of each 'reed' record the response. The instrument obviously has a limited applicability, but nevertheless may be used in the field or in situations where a sophisticated analysis is unnecessary.

The overall response of a model structure may be obtained as an alternative to the response spectrum technique. If the complete time-history of a shock has been recorded in a suitable form, it may be used as the input to excite a model structure in a laboratory environment. This procedure would be feasible if a simple representation of a structure was unrealistic and the structure was subjected to different shock motions.

6.12   Random vibration analysis

A comprehensive consideration of the analysis of random vibrations is beyond the scope of this book. However, it is considered worth while to discuss some aspects of random data analysis in order to provide some continuity in this chapter.

The analysis of random vibration data is carried out using power spectral techniques. The power spectrum of a random vibration quantity provides an indication of the harmonic frequency content of the quantity. It enables an assessment to be made of the relative excitation of different modes of structural vibration provided the power spectrum of the source of vibration remains constant. In such a case, the random vibration is said to be a stationary process. Unfortunately, in most civil engineering situations the power spectrum is not constant and the analysis is more complex. Figure 6.16 illustrates a typical power spectrum plot.

A power spectrum may be obtained by converting the random excitation into an electrical signal and processing the signal with a wave analyzer. The analyzer selects one frequency component of the random quantity to be amplified and measured. There is usually, however, a small contribution from adjacent frequencies. The analysis is conducted over a range of frequencies and the output from the wave analyzer is squared to be consistent with the definition of the power spectrum.

An alternative and more exact method of obtaining the power spectrum is to employ Fourier transform techniques. In this case the spectrum is obtained mathematically via an analogue and digital computation.

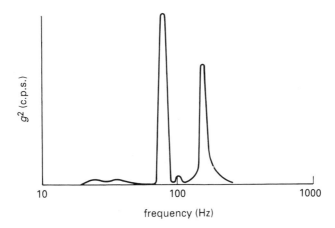

Fig. 6.16   A typical power spectrum plot.

6.13   Vibration or impedance testing

6.13.1   Introduction

Vibration testing and experimentation are an essential part of dynamic analysis. However, there are many practical difficulties associated with the full-scale dynamic testing of civil engineering structures, not least being the size of the structures and the often less than ideal environments in which the structures are situated. Large structures may require extensive instrumentation and may be situated in environments that impose special demands upon equipment. It may also be possible to exert a degree of control over the source of vibration causing the motion.

Despite the difficulties associated with field vibration testing, multi-storey buildings have been instrumented and their response to earthquake motion obtained. The measurement of the response of offshore oil platforms to wave motion is a further example of dynamic field testing.

Modern methods of laboratory structural testing involve the use of model structures. In this case both the input and the structural response are carefully monitored and such experiments are referred to as impedance tests.

The foregoing sections of this chapter have been concerned with a description of the basic aspects of vibration instrumentation, vibration generation and methods of data analysis. It remains to discuss the important details of vibration testing, methodology and interpretation using instrumentation and methods of analysis previously described.

6.13.2   Impedance testing

Recent techniques of experimental dynamics have adopted the principles of impedance testing.

Many of the problems associated with vibration may be described as steady-state and periodic. Consequently, analytical methods have been developed which are based upon steady-state behaviour. Even problems which involve shock or random vibration may be studied using methods which are an extension of periodic methods of analysis.

The basis of impedance testing is the establishment of a ratio which relates a periodic excitation to the resulting periodic response of the structure under test. If a structure is subjected to a periodic force, it will vibrate at the same frequency as the force. However, the response may not necessarily be of the same form as the input motion, i.e. the response to a perfect square wave input will not be a perfect square wave. Moreover, the relationship between force and response is likely to be complex.

For the special case of the harmonic excitation of a linear system and the resulting harmonic response, the relationship between excitation and response is described by the ratio of the amplitudes of the two harmonically varying quantities and the phase angle between them. Both these parameters vary with the frequency of the harmonic vibration and the information they provide is generally referred to as mechanical impedance data. There are a number of different parameters which describe the various specific forms of impedance data and these will be discussed later.

In addition to providing information on natural frequencies, mode shapes and specific damping properties, recent developments in impedance testing provide a means of

(1)   Checking theoretically predicted impedance data to verify an analytical technique and
(2)   Formulating a mathematical model of a structure for further analysis.

These two items require further explanation since they represent the most recent developments and provide scope for further work. With regard to item (1), the comparison of impedance data is a natural development of relating experimental and theoretical natural frequencies. This comparison provides a more precise indication of the behaviour of a theoretical model. A particular problem in the analysis of complex structures is to decide which motion or co-ordinates should be included in the analysis and which should be omitted. The measurement of the appropriate impedance data helps to resolve this problem.

It is item (2) which is most likely to provide a basis for future developments. If an accurate mathematical model of a system has been established, there are a number of uses to which it can be put and some of the more important of these are now mentioned.

The behaviour of the model may be studied when it is subjected to a variety of input motions representing a wide range of excitation. The mathematical model may be modified to represent structural changes, hence obviating the need to make many expensive modifications to the physical laboratory models. A structure may consist of a number of connected parts some of which may be easily analyzed, such as the springs in a vehicle suspension system. Others require an impedance test to determine the appropriate mathematical model, such as the road/tyre interface of the same system. A final dynamic analysis would be conducted using a mathematical model.

6.13.2.1   Basic theory

The basic theory of impedance testing may be discussed using Fig. 6.1. For the present purpose it is assumed that the base of the mass-spring-damper system is rigid and that a harmonic force, p, of frequency ω is considered to be applied to the mass. The general equation motion for this system is

$$m\ddot{x} + c\dot{x} + kx = p \tag{6.9}$$

If $|p|$ is the magnitude of the applied force and ϕ is its phase angle relative to datum, the applied force may be written as

$$p = \bar{p}e^{i\omega t} \tag{6.10}$$

where $\bar{p}$ is a complex quantity related to both the magnitude of the harmonic force $|p|$ and its phase ϕ, thus

$$p = \bar{p}e^{i\omega t} = |p|e^{i(\omega t + \phi_p)} \tag{6.11}$$

It is assumed that the system responds with a harmonic displacement x at the same frequency ω, but with an amplitude $|x|$ and a phase angle ϕ_x relative to the same datum. Therefore

$$x = \bar{x}e^{i\omega t} = |x|e^{i(\omega t + \phi_x)} \tag{6.12}$$

where $\bar{x}$ is a complex quantity.

Substituting for p and x in the equation of motion (6.9) gives

$$(-m\omega^2 + i\omega c + k)\bar{x}e^{i\omega t} = \bar{p}e^{i\omega t} \tag{6.13}$$

$$\frac{\bar{x}}{\bar{p}} = \alpha = \frac{1}{(k - \omega^2 m + i\omega c)} \tag{6.14}$$

If the analysis were to be made for hysteretic or structural damping instead of viscous damping, a similar expression would be obtained. In this case it is convenient to describe the stiffness and damping by a complex stiffness, $k^* = k(1 + i\eta)$, where k is the static stiffness and η the hysteretic damping factor. Equation (6.14) becomes

$$\frac{\bar{x}}{\bar{p}} = \alpha = \frac{1}{(k - \omega^2 m + ik\eta)} \qquad (6.15)$$

The complex quantity α defines the relationship between the harmonic response and the harmonic force and is usually referred to as the receptance. The receptance defines the amplitude ratio and the phase angle between two functions, i.e.

$$\alpha = \frac{\bar{x}}{\bar{p}} = \frac{|x| \, e^{i\phi_x}}{|p| \, e^{i\phi_p}} = \frac{|x|}{|p|} \, e^{i(\phi_x - \phi_p)} = |\alpha| \, e^{i\phi_\alpha} \qquad (6.16)$$

Therefore $|\alpha|$ is the amplitude ratio of displacement/force and ϕ_α is the phase angle between displacement and force.

The three variables which are associated with impedance data may be plotted as graphs of modulus against frequency and phase against frequency. A single polar plot of modulus against phase is an alternative form of data presentation.

Figure 6.17 illustrates the receptance of the single degree of freedom system under consideration for the particular values shown. The logarithmic

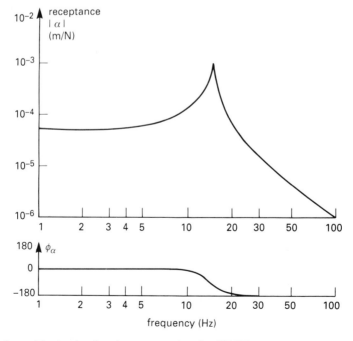

Fig. 6.17 Logarithmic plot for the receptance of a SDOF system.

plot provides a clearer representation of the response of the system and enables the impedance properties of a simple mass or a simple spring to be plotted as straight lines. For the case of a simple mass, m, the equation of motion reduces to

$$m\ddot{x} = p \tag{6.17}$$

and for harmonic motion

$$-\omega^2 m x = p \tag{6.18}$$

hence

$$\alpha_m = \frac{x}{f} = -\frac{1}{\omega^2 m} \tag{6.19}$$

or

$$|\alpha_m| = \frac{1}{\omega^2 m} \quad \text{and} \quad \phi_{am} = 180° \tag{6.20}$$

From the above

$$\log\left(|\alpha_m|\right) = -\log\left(m\right) - 2\log\left(\omega\right) \tag{6.21}$$

which is a straight line having a slope of minus 2 on a plot of $\log\left(|\alpha|\right)$ against $\log\left(\omega\right)$. A similar result is obtained for a simple spring and therefore lines of constant mass and constant stiffness may be superimposed on a logarithmic receptance plot (Fig. 6.18).

The importance of this representation can be seen from this figure in which the receptance curve is almost a straight line at low frequencies,

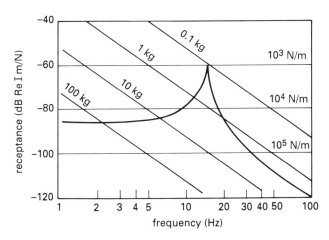

Fig. 6.18   Logarithmic plot of receptance with mass and stiffness lines.

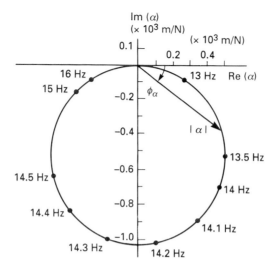

Fig. 6.19 Polar plot for the receptance of a SDOF system.

corresponding to a stiffness line of 2×10^4 N/m, while at high frequencies it corresponds to a mass line of 2.5 kg.

If it is required to examine a narrow frequency band in the vicinity of resonance to obtain say damping characteristics, a polar plot of α against ϕ_α may be more convenient to use than a logarithmic plot (Fig. 6.19).

Equations of motion are usually formulated and solved in terms of the displacement and for this reason the receptance ratio is in wide use. However, it is possible to use the alternative ratios of velocity/force or acceleration/ force. Indeed, the inverse of these ratios may also be used to represent impedance data.

6.13.2.2 *Instrumentation and experimental procedure*

The most convenient method of measuring impedance data involves subjecting a test structure to a sinusoidal force and measuring the sinusoidal response. This procedure is conducted over a range of frequencies. Figure 6.20 shows a typical layout of the instruments required for the measurement of impedance.

The oscillator provides the signal at the desired frequency to the vibration generator, via the power amplifier. The generator subjects the test structure to a mechanical force which is measured by a force transducer. A force transducer is usually a piezo-electric crystal which has been calibrated to measure force. The response of the test structure is measured by piezo-electric accelerometers. During the test, the transducer signals are amplified before passing into the analyzer, which determines the amplitude and phase of each signal relative to the oscillator signal.

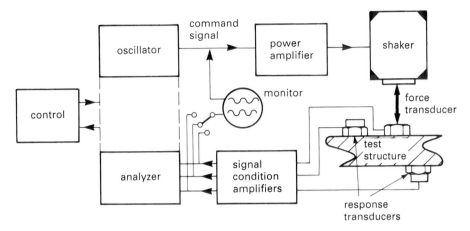

Fig. 6.20 Arrangement of equipment for the measurement of impedance.

In some experimental arrangements it is convenient to combine both force and acceleration-measuring transducers into a single unit which is then called an impedance head.

6.13.2.3 *General comments on the use of impedance data*

It is possible to obtain for a structure with a number of degrees of freedom the distribution of mass and stiffness in the structure.

Impedance curves enable the natural frequency and the damping of each mode to be determined as well as the shape of each mode.

Ewins (1975 & 1976) provides an introduction to impedance testing and gives many further useful references.

6.14 Model testing

The previous sections in this chapter have been concerned with instrumentation and data analysis techniques. These would be appropriate to both full-size and structural model testing. In the full-size case, the engineer will be able to test the structure directly to obtain the relevant dynamic information. However, it may be more desirable to test a model structure, in which case certain conditions must be satisfied. Consideration has not yet been given to the structures for which the foregoing instrumentation and data analysis techniques are relevant. To conclude this chapter, it would seem appropriate to discuss briefly the main principles of dynamic model testing to augment the previous sections. It will be assumed that the reader has some knowledge of dimensional analysis.

6.14.1 Dynamic similarity

The variables for a dynamic problem are the force, f, mass, m, length, l, and time, t. From the theory of dimensional analysis these variables, with the usual notation, may be grouped into a dimensional factor or π factor which is given by

$$\pi = ft^2/ml \tag{6.22}$$

If this single π factor were to be made equal for the model and the prototype, dynamic similarity would be obtained, i.e.

$$\frac{f_m t_m^2}{m_m l_m} = \frac{f_p t_p^2}{m_p l_p} \tag{6.23}$$

where the subscripts m and p relate to the model and the prototype respectively. From this expression, the following scale factors would be defined

$$\frac{f_p}{f_m} = \lambda_f ; \quad \frac{t_p}{t_m} = \lambda_t ; \quad \frac{m_p}{m_t} = \lambda_m ; \quad \frac{l_p}{l_m} = \lambda_l \tag{6.24}$$

and the model/prototype relation becomes

$$\lambda_f = \lambda_l \lambda_m / \lambda_t^2 \tag{6.25}$$

This equation which relates the scale factors is often referred to as an expression of the principle of dynamic similarity.

6.14.2 Model design based on system variables

In certain cases it is not possible to satisfy completely the requirement of dynamic similarity. It is then usual to group the problem variables into dimensionless parameters, which are required to be identical for the prototype and the model. For example, the grouping for a structure subjected to dynamic loading would be given by

$$\phi \left(\frac{\omega^2 l^2 \rho}{E}, \frac{\sigma l^2}{f}, \frac{\rho g l}{E}, \frac{f}{E l^2}, \frac{l \omega^2}{g}, \frac{a_n}{l}, \frac{\delta_n}{l}, \frac{c_n}{f}, v \right) \tag{6.26}$$

where

σ = stress E = modulus of elasticity
a_n = geometric dimensions l = length
ρ = density δ_n = deflections
f = reference load g_n = gravitational acceleration
c_n = loads ω = frequency
v = Poisson's ratio

It is not often that all these groupings must be satisfied simultaneously. In most problems there are special conditions which permit a relaxation of the need for equality of all groups.

6.14.3 Model design based on system equations of motion

If the system equation of the model and the prototype are known, a simpler approach may be adopted with some restrictions on the modelling conditions removed. This is achieved by grouping all the quantities in the equation into dimensionless parameters, which leads to a system model. A simple example will be used to illustrate the idea. Consider the equation of motion for the lateral vibration of a uniform beam of length l which is fixed at one end and free at the other

$$EI\frac{\partial^4 y}{\partial x^4} + \rho A\frac{\partial^2 y}{\partial t^2} = 0 \tag{6.27}$$

where

y = deflection ρ = mass density
E = modulus of elasticity A = cross-sectional area
I = second moment of area x = an arbitrary distance along the beam
t = time

The boundary conditions are

at $x = 0$ $\partial^2 y/\partial x^2 = 0$ $\partial^3 y/\partial x^3 = 0$
at $x = 0$ $y = 0$ $\partial y/\partial x = 0$ $\qquad\qquad$ (6.28)

The dimensionless parameters may be taken as

$$\alpha = x/l \qquad \beta = y/l \qquad \tau = ct/l \qquad \xi = I/Al^2 \tag{6.29}$$

where $c = \sqrt{(E/\rho)}$, the velocity of wave propagation. With these new variables Equation (6.27) takes the form

$$\frac{\xi\partial^4\beta}{\partial\alpha^4} + \frac{\partial^2\beta}{\partial\tau^2} = 0 \tag{6.30}$$

and the boundary conditions are approximately modified. The necessary condition for model design is $(\xi)_m = (\xi)_p$ and complete geometric similarity is unnecessary.

References

ACI Committee 442 (1988) Response of concrete buildings to lateral forces. ACI 442R-88, Feb 88.

Alpan, I. & Meidav, T.S. (1963) The effect of pile driving on adjacent buildings, a case history. *RILEM Symp, Budapest*, **2**, 171.

Applied Technical Council (ATC) (1978) *Tentative Provisions for the Development of Seismic Regulations for Buildings*. ATC-3-06, National Bureau of Standards, USA, SP 510.

Ashley, C. (1978) Proposed international standards concerning vibration in buildings. *Conf on Instrumentation for Ground Vibration and Earthquakes*. ICE, pp. 153–62.

Barkan, D.D. (1962) *Dynamics of Bases and Foundations*. McGraw-Hill, London and New York.

Bean, R. & Page, J. (1976) *Traffic-induced Ground Vibration in the Vicinity of Road Tunnels*. TRRL supplementary report 218, Transport & Road Research Laboratory, Crowthorne, UK.

Blanchard, J. *et al.* (1977) Design criteria and analysis for dynamic loading of footbridges. *Symp on Dynamic Behaviour of Bridges*, Paper 7. TRRL supplementary report 275, Transport & Road Reseach Laboratory, Crowthorne, UK.

Buzdugan, Gh. (1968) Fatigue coefficient in machine foundation design. *Proc 3rd Conf on Dimensioning*, Budapest, 1968, 61.

British Standards Institution (1972) CP3: *Code of basic data for the design of buildings*. Chapter V: Part 2, *Wind loads*. BSI, London.

British Standards Institution (1976 & 1978) BS 4675: *Mechanical vibration in rotating machinery*, Part 1: 1976; Part 2: 1978. BSI, London.

British Standards Institution (1984) BS 6472: *Guide to evaluation of human exposure to vibration in buildings (1 Hz to 80 Hz)*. BSI, London.

British Standards Institution (1985) BS 6611: *Guide to evaluation of the response of occupants of fixed structures, especially buildings and offshore structures, to low-frequency horizontal motion (0.063 Hz to 1 Hz)*. BSI, London.

Chen, P.W. & Robertson, L.E. (1972) Human perception thresholds of horizontal motion. *J Struct Div, ASCE*, **98**, ST8, 1681.

Ciesielski, R. (1963) Charts for evaluating the effect of externally applied vibrations and impacts on masonry constructions. *RILEM Symp*, Budapest, **1**, 19.

Clough, R.W. & Penzien, J. (1975) *Dynamics of Structures*. McGraw-Hill, New York.

Crandell, E.J. (1949) Ground vibration due to blast loading and its effects on structures. *J Boston Soc Civ Engrs*, **36**(2), 222.

Crockett, J.H.A. (1963) Traffic vibration damage in medieval cathedrals. *RILEM Symp*, Budapest, **1**, 303.

Dawance, G. (1957) Measurements and effects of vibration in houses and industrial buildings. *Ann Inst Batim*, **10**, 115−6, 713.

DEGEBO (1936) *The Application of Dynamic Investigations of Building Grounds*. Ministry of Works Library translation no 81.

Den Hartog, J.P. (1947) *Mechanical Vibrations*, 3rd edn. McGraw-Hill, New York.

Department of Energy (DoE) (1990) *Offshore Installations: Guidance on Design, Construction and Certification*. HMSO, London.

Deutsche Industrie Norm (1939) *Erschütterungsschutz im Beuwesen*. DIN, Berlin, DIN 4150.

Deutsche Industrie Norm (1958) *Fundamente für Amboss − (Schabotte-) Hämmer*. DIN, Berlin, DIN 4025.

Deutsche Industrie Norm (1986) *Erschütterungsschutz im Beuwesen*. DIN, Berlin, DIN, 4150, Part 2.

Dieckmann, D. (1958) A study of the influence of vibration on man. *Ergonomics*, **1**(4), 347.

Edwards, A.T. & Northwood, T.D. (1960) Experimental studies of effects of blasting on structures. *Engineering (London)*, **210**, 538.

Ewins, D.J. (1975 & 1976) Measurement and application of mechanical impedance data. *J Soc Environ Engrs*, Dec 1975, Mar & Jun 1976.

Grootenhuis, P. (1967) The attenuation of noise and ground vibrations from railways. *Symp Soc Environ Engrs*, 57.

Hansen, R.J. *et al.* (1973) Human response to wind induced motion of buildings. *J. Struct Div, ASCE*, **99**, ST7, 1589−605.

Harris, C.M. & Crede, C.E. (Eds) (1987) *Shock and Vibration Handbook*, Vols 1−3. McGraw-Hill.

Irwin, A.W. (1978) Human response to dynamic motion of structures. *The Structural Engineer*, **56A**, 237−44.

Irwin, A.W. (1981) Perception, comfort and performance criteria for human beings exposed to whole body pure yaw vibration, continuing yaw and translational components. *J Sound & Vibration*, **76**, 481.

Koch, H.W. (1953) Ermittlung der Wirkung von Banwerksschwingurgen. *VDI-Zeitschr*, **95**, 733.

Kolsky, H. (1964) *Stress Waves in Solids*. Dover Press, New York.

Korchinski, I.L. (1948) *Computation of Building Structures for Vibration Load*. Moscow.

Langefors, U. & Kiehlstrom, B. (1963) *Rock Blasting*, p. 405. Almquist & Wibsell. Wiley, Stockholm and New York.

Lehigh University (1972) *Proc Intl Conf on Tall Buildings*, Aug.

Leonard, D.R. (1966) *Human Tolerance for Bridge Vibrations*. TRRL report

LR34, Transport & Road Research Laboratory, Crowthorne, UK.

Major, A. (1962) *Vibration Analysis and Design of Foundations for Machines and Turbines.* Collets, London.

Major, A. (1980) *Dynamics in Civil Engineering.* Collets, London.

Martin, D.J. (1978) *Low Frequency Traffic Noise and Building Vibration.* TRRL supplementary report 429, Transport & Road Research Laboratory, Crowthorne, UK.

Matsumoto, Y. *et al.* (1978) Dynamic design of footbridges. *Proc Intl Assoc Bridge & Struct Engng,* P-17/78, Aug, 1–15.

Morton, B.A. (1967) Building on rubber. *Building Tech & Management,* 6 June.

Newmark, N.M. & Hansen, R.J. (1961) Design of blast-resistant structures. In *Shock and Vibration Handbook,* Vol 3, pp. 49–51. McGraw-Hill, New Yok.

Newmark, N.M. *et al.* (1973) Seismic design spectra for nuclear power plants. *J Power Div, ASCE,* Nov, 287.

Newmark, N.M. (1978) Seismic design criteria for structures and facilities of the trans-Alaska pipeline systems. *Proc US National Conf on Earthquake Engng EER,* Oakland, 1975.

Nicholls, H.R. *et al.* (1971) *Blasting Vibrations and their Effects on Structures.* US Bureau of Mines, Bulletin 656.

Novak, M. & Tanaka, H. (1974) Effect of turbulence on galloping instability. *Proc ASCE,* EM1, 27.

Oehler, L.T. (1957) Vibration susceptibilities of various high bridge types. *Proc ASCE,* **83**, ST4, 1318.

Postlethwaite, F. (1944) Human susceptibility to vibration. *Engineering (London),* **157**, 61.

Pretlove, A.J. (1966) Some current methods in vibration measurement. In *Vibration in Civil Engineering* (Ed by B.O. Skipp). Butterworths, London.

Rathbone, T.C. (1939) Vibration tolerance. *Power Plant Engng,* Nov.

Rausch, E. (1959) *Machinery Foundations and Dynamic Problems in Buildings.* VDI Verlag, Dusseldorf.

Reiher, H.J. & Meister, F.J. (1931) Human sensitivity to vibration. *Forsch. auf. dem. Geb. des. Ing.,* **2**, 11, 381. (Translated Rep. F-TS-616-RE, Wright Field, 1946.)

Richter, C.F. (1935) An instrumental earthquake scale. *Bull Seis Soc Am,* **25**, 1.

Richter, C.F. (1958) *Elementary Seismology.* W.H. Freeman & Co, San Francisco, USA.

Rosivall, F. & Goschy, B. (1963) Dynamic effects of the road traffic on the building of the Palace of Arts of Budapest. *RILEM Symp,* Budapest, **1**, 375.

Simiu, E. & Scanlan, R.H. (1986) *Wind Effects on Structures: an Introduction to Wind Engineering.* Wiley Interscience, New York.

Sisking, D.E. *et al.* (1980) *Structure Response and Damage Produced by Ground Vibration from Surface Mining.* US Bureau of Mines, Report of investigations 8507.

Skipp, B.O. (Ed) (1966) *Vibration in Civil Engineering.* Butterworths, London.

Smith, E.A. (1962) Pile-driving analysis by the wave equation. *Trans ASCE,* **127**, part 1, 1145.

Smith, J.W. (1988) *Vibration of Structures: Application in Civil Engineering Design.* Chapman & Hall.

Soliman, J.I. (1963) Criteria for permissible levels of industrial vibrations with regard to their effects on human beings and buildings. *RILEM Symp*, Budapest, **1**, 111.

Splittgerber, H. (1978) Effects of vibrations on buildings and on occupants of buildings. *Conf on Instrumentation for Ground Vibration and Earthquakes*, pp. 147–52. Institution of Civil Engineers, London.

Srinivasulau, P. & Vaidyanathan, C.V. (1976) *Handbook of Machine Foundations*. Tata McGraw-Hill, New Delhi, India.

Steffens, R.J. (1952) The assessment of vibration intensity and its application to the study of building vibrations. *Mat Build Stud Spec*. Report no 19, HMSO, London.

Steffens, R.J. (1966) Some aspects of structural vibration. In *Vibrations in Civil Engineering* (Ed by B.O. Skipp). Butterworths, London.

Structural Engineers' Association (SEA) (1973) *Recommended Lateral Force Requirements and Commentary*. Seismology Committee, SEA, California, USA.

Teichmann, G.A. & Westwater, R. (1957) Blasting and associated vibrations. *Engineering (London)*, **183**, 460.

The Blasters' Handbook (1969) Du Pont de Nemours, Wilmington.

Uniform Building Code (UBC) (1982) *Intl Conf of Building Officials*, Whittier, California, USA.

US Bureau of Mines (1936) *Earth Vibrations Caused by Quarry Blasts*. RI 3353.

US Bureau of Mines (1937) *Earth Vibrations Caused by Mine Blasting*. RI 3407.

US Bureau of Mines (1938a) *Earth Vibrations Caused by Ground Vibration*. RI 3431.

US Bureau of Mines (1938b) *House Movement Induced by Mechanical Agitation and Quarry Blasting*. RI 3431.

US Bureau of Mines (1940) *House Movement Induced by Mechanical Agitation and Quarry Blasting*. RI 3542.

US Bureau of Mines (1942) *Seismic Effect of Quarry Blasting*. Bulletin 442.

Verein Deutscher Ingenieure (VDI) (1941) *Richtlinen, Gestaltung und Anwendung van Gummiteilen*.

Waller, R.A. (1965) A block of flats isolated from vibration. *Symp Soc Environ Engrs*, 1965.

Waller, R.A. (1966) Some economic aspects of anti-vibration devices. *Proc Symp Vibration in Civ Engng*. Butterworths, London.

Waller, R.A. (1970) Design techniques in practical wind problems. In *Dynamic Waves in Civil Engineering*, p. 57. Wiley Interscience, London.

Warbarton, G.B. (1976) *The Dynamic Behaviour of Structures*, 2nd edn. Pergamon, Oxford.

Zeller, W. (1931) Stärkebestimmung von mechanischen Erschütterungen. *Bauing 1931*, **12**, 32/33, 586.

Zeller, W. (1933) Proposal for a measure of the strength of vibration. *VDI-Zeitschr*, **77**, 323.

Zeller, W. (1941) Bedentung der Dämpfung fur die Erschüterungs dämmung bei Hammeranlagen. *VDI-Zeitschr*, **85**, 348.

Zeller, W. (1949) Units of measurement for strength and sensitivity of vibrations. *Automob − tech Zeitschr*, **51**, 95.

Index